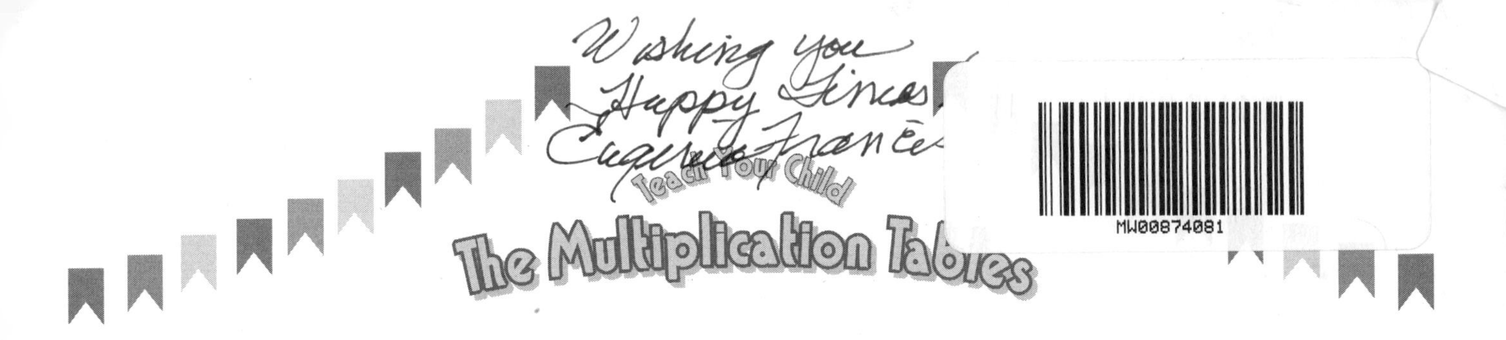

The recognition of patterns is a creative way to have students develop understanding for the concept of multiplication. Pattern analysis should be part of the elementary study in mathematics as it is also viewed as foundational skills for algebraic reasoning.

Michael Kestner
Mathematics and Science Partnership Program
Office of Elementary and Secondary Education/United States Department of Education.

Teach Your Child the Multiplication Tables is everything I think a multiplication workbook should be, and my daughter thinks it's lots of fun. Set around a circus theme, the book is child-friendly with lots of pictures, large print, and plenty of space for writing the answers. The best part of this book, however, is that rote memorization is not the goal, rather understanding the inherent patterns is.

My daughter works through this workbook at her own pace, whenever she's in the mood for it, and she's already had quite a few "ah-ha!" moments as she recognizes and predicts the various patterns.

Ruth Pell
California Homeschool News

Many children struggle learning and recalling multiplication facts, and need other techniques rather than rote memorization to master these skills. Eugenia Francis' workbook utilizes wonderful, brain-compatible strategies and methods to do so - such as learning to recognize and attend to patterns for each of the multiplication tables, using memory tricks/mnemonics, and other engaging and fun techniques. I recommend **Teach Your Child the Multiplication Tables** as a helpful resource for children to learn the math facts and understand the principles of multiplication.

Sandra Rief, author of **How to Reach and Teach Children with ADD/ADHD** and **How to Reach and Teach Children in the Inclusive Classroom**

Teach Your Child the Multiplication Tables is a wonderfully entertaining, clear, and memorable way to understand multiplication and to learn the times tables! Younger students will also benefit from this book; cute characters and cartoon creatures appear on each page. As soon as the book arrived, my 5 year old daughter was eager to work on it. **Teach Your Child the Multiplication Tables** is one of my top picks for 3rd graders at www.mathmom.com.

Linda Burks

Articles on Eugenia Francis and her workbook have appeared in **The Wall Street Journal** and **Education Matters**. In an interview with **Home Education Magazine**, the author discusses the benefits of her method for children with special needs.

When you finish this workbook,

you'll know your multiplication tables!

Ringmaster Rudy

Teach Your Child

The Multiplication Tables

Fast, Fun & Easy!

with Dazzling Patterns, Grids and Tricks!

TeaCHildMath™

Eugenia Francis

TeaCHildMath M. R.

TeaCHildMath.com

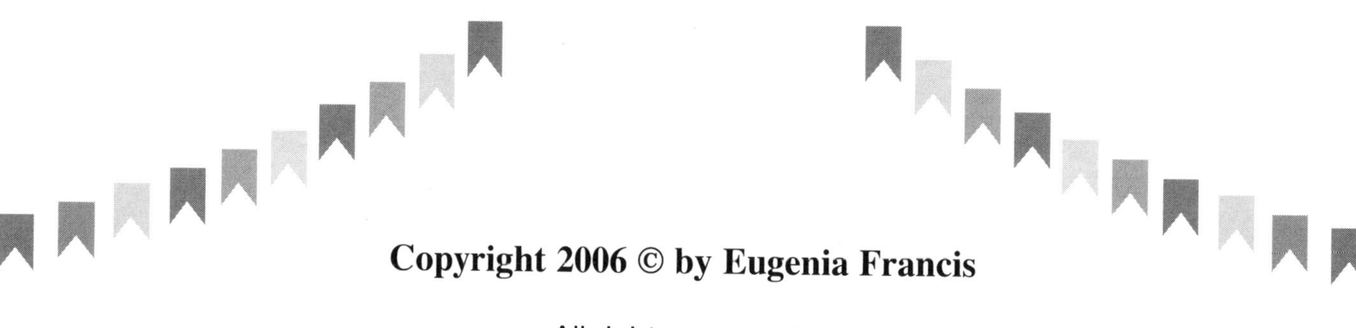

To contact the author visit www.TeaChildMath.com and enter message on the Contact page.

Special thanks to illustrators Michael Likens and Rudy Rodriguez at Gopixel Design Studios, the Intellectual Property law firm of Knobbe Martens and, of course, my son, Scott Francis, who inspired this journey.

Printed in the United States of America

Charleston, SC

Second edition March 2013

ISBN-13: 978-1482079470

ISBN-10: 148207947X

Preface to the Second Edition

This second edition of **Teach Your Child the Multiplication Tables** incorporates suggestions made by parents, teachers and students. I am most grateful for their input. My goal is multiplication mastery for all children.

This workbook can be used as a primary or supplemental text in the classroom, homeschool or home by:

- a parent, teacher, or learning specialist working one-on-one with a student.
- a teacher with a classroom of students.

Multiplication is an essential building block of mathematics. A student who has not mastered the times tables has difficulty succeeding in mathematics beyond the third grade. Division, fractions, percentages, decimals and algebra - all require a solid foundation in multiplication. Standardized tests are timed. Word problems become more complex. Students cannot get bogged down trying to figure out the answer to 6 times 8 or other such basic multiplication facts. The answer must be retrieved from long-term memory with speed, freeing up the working memory for problem solving. Recall needs to be automatic.

Traditionally, students have struggled to learn the tables through rote memorization. Rote memorization is passive and mechanical. Students typically find it boring. There is a better way. My method is based on patterns. Discovering patterns is active, creative and engaging. According to Michael Kestner of the Mathematics and Science Partnership Program of the Department of Education in a letter to the author regarding this workbook:

> The recognition of patterns is a creative way to have students develop understanding for the concept of multiplication. Pattern analysis should be part of the elementary study in mathematics as it is also viewed as foundational skills for algebraic reasoning.

Patterns provide a way to organize. Our brains seem designed to search for patterns. Learning one math fact at a time through rote memorization is highly inefficient. Patterns are efficient because students only have to learn one pattern for the entire table. Special needs children, such as those with ADD/ADHD and dyslexia, can better visualize and recall where a number is placed when they see a pattern. This is true of all children. In addition, children in the autistic spectrum seem to have a natural affinity for patterns.

The TeaCHildMath™ method utilizes both left and right-brain strategies to teach multiplication. There are marked differences between children who are left-hemisphere dominant and those who are right dominant. Whereas the left-hemisphere dominant child can construct the whole from parts, the right dominant prefers the big picture, seeing patterns and making connections. Special needs children are often right-brain dominant. Learning the multiplication tables is easier when both hemispheres are engaged.

The TeaCHildMath™ Method:

Perform a diagnostic (page 167) to determine which tables your child or students may already know. Although students may know a few of the tables, start at the beginning of the workbook so that students learn the underlying concepts such as the **commutative property** of multiplication: 6 x 4 is the same as 4 x 6.

The TeaCHildMath™ method uses a hundred-square grid (Tables 1 -10) to teach each table. It is easier to learn a table when seen in context of the others. Reinforce learning by having students occasionally count the squares on the grid for a multiplication problem. Example: 5 x 5 = ___. Have students count the 25 squares.

Each table has a pattern. Some students need extra help connecting the pattern to the multiplication table. Have the student say the entire table out loud while filling in the pattern. For example, when filling in the 8-6-4-2-0 pattern for table 8, have the student say: "8 times 1 is 8, 8 times 2 is 16, 8 times 3 is 24" and so on.

Instead of teaching tables 1 through 10 sequentially, the TeaCHildMath™ method teaches tables for EVEN numbers which are easy to learn before teaching tables for ODD numbers.

The TeaCHildMath™ Sequence:

1. Tables 1 and 10 and multiplying by 0.
2. Tables for EVEN numbers. Tables 2, 4, 6 and 8 have similar patterns that are easy to learn as they all end in some combination of 2-4-6-8-0. These patterns repeat forever. The skip counting pages (example: page 15) illustrate this point.
3. How to determine whether a number is ODD.
4. The **ODD/EVEN rule of multiplication**. The product of two factors is always an EVEN number unless both factors are ODD. (See page 59.)
5. Tables for ODD numbers. Tables 5, 9, 3 and 7 are presented in order of increasing difficulty. Starting with the easier tables builds confidence. These tables too have intriguing patterns that repeat forever. The skip counting pages (example: page 83) illustrate this point.
6. Diagonal patterns. These illustrate the commutative property of the tables.
7. Double-digit multiplication.
8. Division without a remainder. Division with a remainder.
9. Word problems of increasing difficulty appear throughout. All word problems have visual cues and a real context such as the Circus Snack Menu. Each is broken down step by step to show students how to solve the problem. Learning these problem-solving strategies, students develop math skills.
10. *Magic Circus* skip counting pages, *Color the Clown, Multiplication Bingo, Multiple Mystery* and other fun activities reinforce learning.
11. Each page has engaging artwork. Students can color the circus figures or trace the watermarks on the grids.

Multiplication equations are a bridge to algebra. Most of the multiplication problems in the workbook are written as an equation rather than the traditional vertical multiplication problem:

$$9 \times 8 = \_\_\_\_ \quad \text{vs.} \quad \begin{array}{r} 9 \\ \underline{\times 8} \end{array}$$

By learning to solve multiplication problems written as an equation, students are preparing themselves for algebra.

$9 \times \_\_\_ = 72$ is a step toward $9n = 72$.

Teach Your Child the Multiplication Tables will give students the confidence and skills to advance in math. Not only will they have learned the multiplication tables but also the underlying principles of multiplication.

Since publication of **Teach Your Child the MultiplicationTables,** I have received emails from parents and teachers attesting to the success of my method.

My workbook has been endorsed by mathematicians and learning specialists. In addition, I have been interviewed by **Home Education Magazine**. My story, university English instructor turned children's math book writer, appeared in **The Wall Street Journal**.

Please comment on your child's or classroom's experience on the **Contact Form** on my website: **www.TeaCHildMath.com**.

If you would like to work with me in obtaining a grant testing the efficacy of my method in your classroom, please contact me on the **Contact Form** on my website.

For Learning Aids and other TeaCHildMath™ products, visit: www.TeaCHildMath.com.

Eugenia Francis
Irvine, California

PARENT/TEACHER SURVEY

I appreciate hearing from parents, teachers and students. Please comment on your experience with my workbook in teaching your child or classroom the multiplication tables. Your input will help enrich the learning experience of all children.

Log on to **www.TeaCHildMath.com** to share your comments on my blog or on **Contact Us**.

The following editions are available:

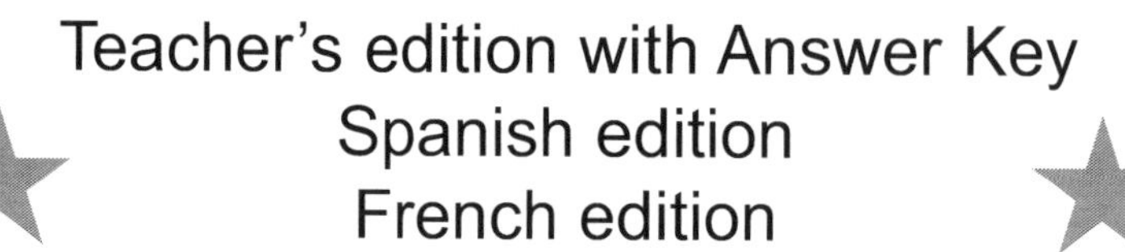

Teacher's edition with Answer Key
Spanish edition
French edition

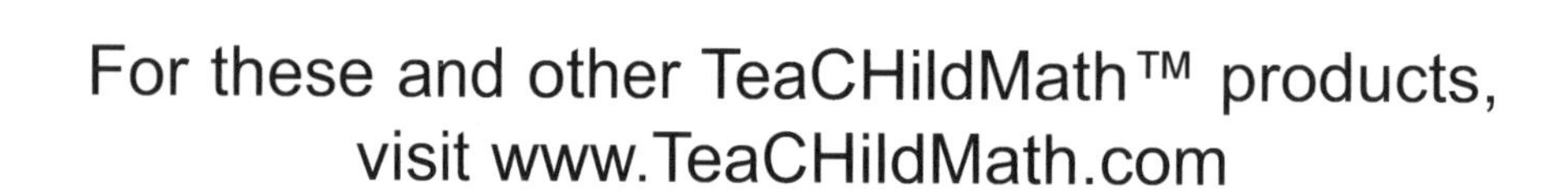

For these and other TeaCHildMath™ products,
visit www.TeaCHildMath.com

Table of Contents

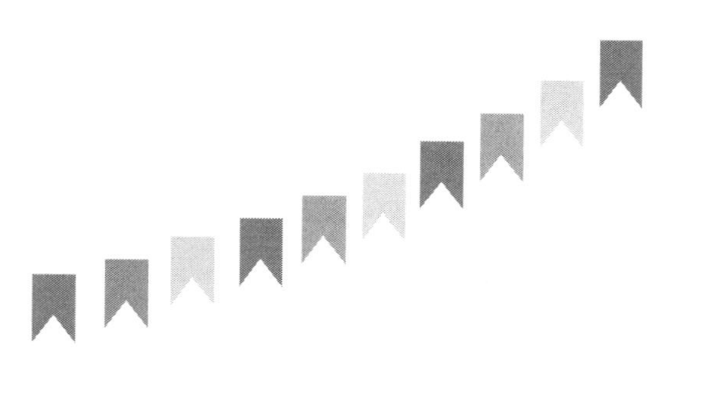

Included in the above are review sheets to reinforce learning & evaluate progress.

Introduction

Like most third graders, my son found learning the multiplication tables a challenge. As Scott struggled with endless drills, I realized there had to be a better way than rote memorization!

At the kitchen table, I drew a grid for tables 1-10. Not surprisingly, Scott filled in tables 1, 2, 5 and 10, the tables with easy patterns. I remembered a trick for table 9. On the grid, I had Scott pencil in 0-9 down the 9 column and then 9-0 to the right of these numbers. "There you know the 9's!" I said. Scott was amazed by this trick. Table 9, he decided, was super easy!

Well, then, why not find easy patterns or tricks for the rest of the tables? Tables for 2, 4, 6, and 8, we discovered, end in some combination of 2-4-6-8-0. Tables for odd numbers also have disctinct patterns. Patterns made Scott smile. He could see the structure and knew he got it right.

What other patterns could we discover such as whether the multiple would be odd or even? Odd multiples, we found, were few in number. Why? Because an even number times ANY number (odd or even) is EVEN. An odd multiple results ONLY when an ODD number multiplies another ODD. Each pattern boosted Scott's confidence. Our fridge eventually was papered with amazing patterns. Patterns appeal to children. Learning to recognize these teaches analytical skills.

Each time Scott filled in the pattern for a new table on a grid for tables 1-10, I had him fill in other tables he had learned. Rather than learning each table in isolation, he learned each in context of the others. At the same time, he learned the commutative property of multiplication: 8 x 6 is the same as 6 x 8. As he filled in each table, I had him connect the pattern to the table. For example, while filling in the 8-6-4-2-0 pattern for table 8, I had him repeat: "8 times 1 is 8, 8 times 2 is 16, 8 times 3 is 24" and so on. Scott learned by doing.

Like many children, Scott was a visual learner. My method helped him discover patterns, integrate the concept of multiplication and actually made learning the times tables fun.

Pattern recognition greatly benefits children with special needs. They can better recall and visualize the table when they see a pattern. This is true for all children. Patterns aid memory.

A college English instructor, I was trained to discover patterns and spot key differences. Patterns whether in literature or math reveal the underlying structure. There is an inherent simplicity in them, an inherent beauty. Math should engage your child's imagination.

Patterns introduce children to the magic of math. As one reviewer said, “The essence of mathematics is patterns. **Teach Your Child the Multiplication Tables** enables children to experience the joy of discovering mathematical patterns at an early age.” (Charles Curto, Irvine, California)

Teachers, why not bring the magic of math into the classroom? If all third graders were to genuinely like math, they would be more likely to succeed in school. Mathematics is more than manipulating numbers. It is an analytical way of thinking that trains children for all academic subjects. A solid foundation in math opens the door to any number of rewarding careers.

My goal in writing **Teach Your Child the Multiplication Tables, Fast, Fun and Easy** was not only times tables mastery for all children but to instill in them a love of numbers and fascination with math. The question I am asked most often is: “Why did you, a college English instructor, write a children’s math book?”

My reply is: “As an English instructor, I saw the times tables in a different way and came up with an innovative way to teach them. Why not publish my method and help other families? I decided on a circus theme and hired graphic artists to design fun characters that would appeal to children. I believe if more of us would do for other people’s children what we do for our own, the world would be a better place. I want to help your child as I did mine.”

Please share your child’s or classroom’s experience with my workbook on my blog on my website at **www.TeaCHildMath.com.**

Discovery makes learning fun!

Crazy, Crazy Circus!

It's a wild night at the Magic Circus!
The circus train roared into the big top.
Before Ringmaster Rudy could blow his whistle,
lions jumped out.
Elephants broke out. Giraffes leaped out.
Bears hopped out. Monkeys climbed out.
Clowns tumbled out and ran after them.
The Magic Circus is a crazy, crazy circus!

Ringmaster Rudy needs your help!
He must round up all the lions,
elephants, giraffes, bears, monkeys and clowns!
It will take too long to count!
It will take too long to add!
Lions, elephants, giraffes, bears, monkeys and clowns
have to do their acts!
Let's help Rudy find patterns and multiply!
Here's your ticket to the Magic Circus.
Step right in!

Find a Pattern and Multiply

You can COUNT objects one by one but this is so ***s-l-o-w.***
You can find a pattern and ADD but this is too ***s-l-o-w.***
Or you can find a pattern, MULTIPY and go ***fast!***

Count

24

24 steps
So slow

Add

4
4
4
4
4
+4
24

6 steps
Slow

Multiply

4 x 6 = 24
4 six times = 24

4
x6
24

1 step!
Fast!

Look for patterns. Each appears in a row of 4.

There are 6 rows of .

You can **ADD:** 4+4+4+4+4+4=24 **SLOW**

You can **MULTIPLY:** 4 X 6 = 24 ***FAST!!!!***

Look for PATTERNS

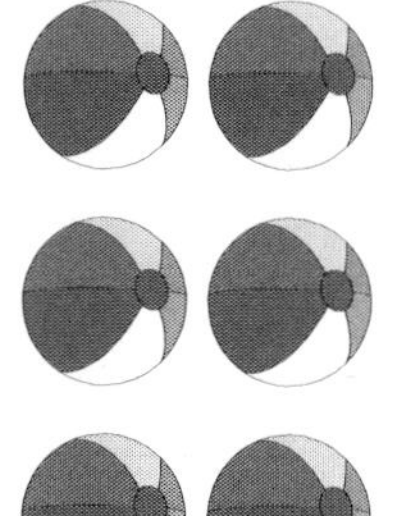

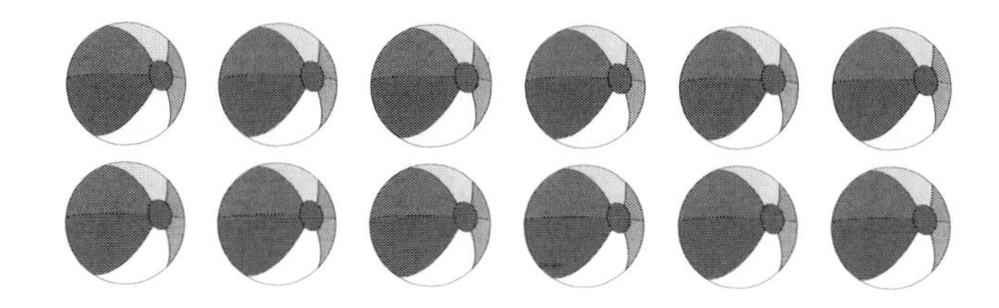

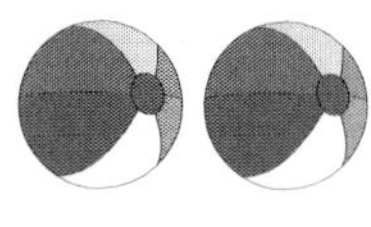

2 six times is the same as **6 two times.**

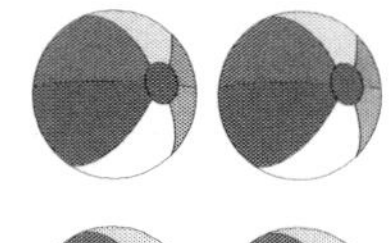

Turn page sideways to check.

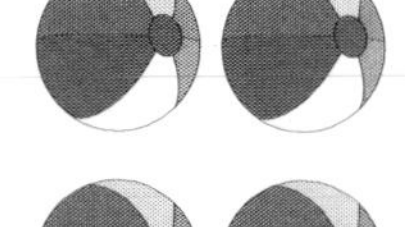

6 x 2 = 12

2 x 6 = 12

Find the pattern:

5 four times?

Turn the page sideways.

4 five times?

Fill in: 5 x _4_ = 20

4 x _5_ = 20

Find the pattern:

9 two times?

2 nine times?

Fill in: 9 x _2_ = 18

2 x _9_ = 18

Find a pattern and multiply:

5 x 3 = 15
3 x 5 = 15

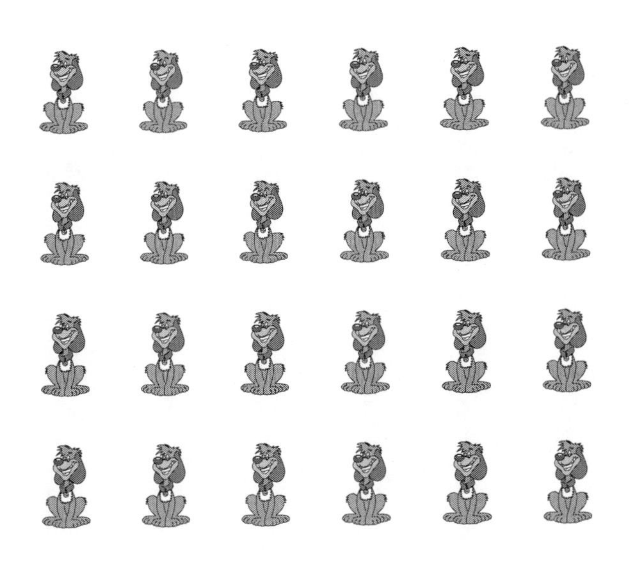

6 x 4 = 24
4 x 6 = 24

7 x 2 = 14
2 x 7 = 14

5 x 1 = 5
1 x 5 = 5

How Do Numbers Multiply?

Let's Multiply the Dots and See

Multiply 2 x 3 on the grid.
What do you discover when you turn the page sideways?

Two dots three times is the same as three dots two times. 2 x 3 = 6 3 X 2 = 6

Now multiply 3 x 4 and 4 x 3.

X	1	2	3	4	5
1	•	••	•••	••••	•••••
2	• •	•• ••	••• •••	•••• ••••	••••• •••••
3	• • •	•• •• ••	••• ••• •••	•••• •••• ••••	••••• ••••• •••••
4	• • • •	•• •• •• ••	••• ••• ••• •••	•••• •••• •••• ••••	••••• ••••• ••••• •••••
5	• • • • •	•• •• •• •• ••	••• ••• ••• ••• •••	•••• •••• •••• •••• ••••	••••• ••••• ••••• ••••• •••••

The trick to the ZERO is: the answer is always ZERO.

1 x 0 = 0
2 x 0 = ____
3 x 0 = ____
4 x 0 = ____
5 x 0 = ____

0 x 1 = 0
0 x 2 = ____
0 x 3 = ____
0 x 4 = ____
0 x 5 = ____

Rudy's Magic Zero Rule

Any number multiplied by 0 is 0.
0 multiplied by any number is 0.

X	0	1	2	3	4	5	6	7	8	9	10
0	0	0	0	0	0	0	0	0	0	0	0
1	0										
2	0										
3	0										
4	0										
5	0										
6	0										
7	0										
8	0										
9	0										
10	0										

Tables 1 and 10 Magic!

Super Easy 1 and 10

Multiplying 1 x 1-10 is **super easy!**
Just number 1-10!
For the 10 x table ADD a 0.
So easy! Fill in the darkened squares.

X	1	2	3	4	5	6	7	8	9	10
1	1									1_
2	2									2_
3	3									3_
4	4									4_
5	5									5_
6	6									6_
7	7									7_
8	8									8_
9	9									9_
10	10									10_

Help Bernice complete the pattern and find her way home to **300 Magic Circus Drive.**

Clown Fun

Fill in the blanks.

10 x ___ = 10

9 x ___ = 9

8 x ___ = 8

7 x ___ = 7

6 x ___ = 6

5 x ___ = 5

4 x ___ = 4

3 x ___ = 3

2 x ___ = 2

1 x ___ = 1

10 x ___ = 0

9 x ___ = 0

8 x ___ = 0

7 x ___ = 0

6 x ___ = 0

5 x ___ = 0

4 x ___ = 0

3 x ___ = 0

2 x ___ = 0

1 x ___ = 0

Circus Snacks

Hot Dogs	$3	Ice Cream	$2
Pretzels	$1	Popcorn	$3
Cotton Candy	$2	Soda	$1

Carla bought 10 hot dogs at the circus for her friends. How much did she spend?

10 x $3 = $30

Jim bought his 5 brothers each a soda. How much did he spend?

5 x $1 = $5

The twins bought 10 bags of popcorn and 5 sodas. How much did they spend?

10 x $3 = $30
5 x $1 = $5
Total: 30 + 5 = $35

Cindy bought 1 ice cream bar and 2 sodas. How much did she spend?

_____ x _____ = $_____
_____ x _____ = $_____
Total: _____ + _____ = $_____

Ricky and Bill bought 10 hot dogs, 7 sodas and 3 pretzels. How much did they spend?

_____ x _____ = $_____
_____ x _____ = $_____
_____ x _____ = $_____
Total: _____ + _____ + _____ = $_____

Lisa bought her family 6 sodas, 5 pretzels and 10 hot dogs. How much did she spend?

_____ x _____ = $_____
_____ x _____ = $_____
_____ x _____ = $_____
Total: _____ + _____ + _____ = $_____

Magic Circus Fun

__ x __ = __ s

__ x __ = __ s

__ x __ = __ s

2-4-6-8-O!

Fill in the column with the 2-4-6-8-0 pattern.
Multiply table 2 across. How easy is that?

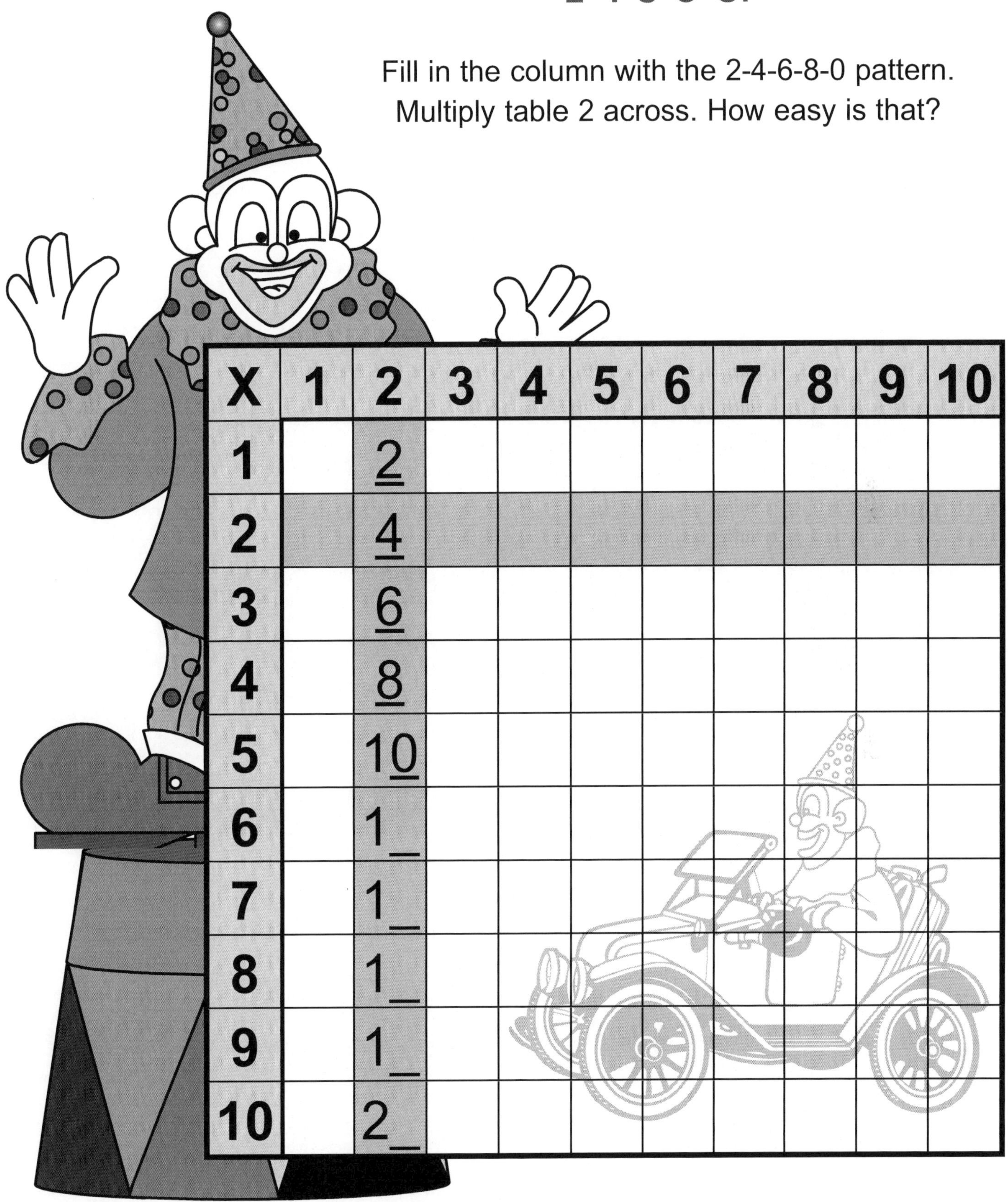

X	1	2	3	4	5	6	7	8	9	10
1		2								
2		4								
3		6								
4		8								
5		10								
6		1_								
7		1_								
8		1_								
9		1_								
10		2_								

Tables 1 and 10 Review

X	1	2	3	4	5	6	7	8	9	10
1										
2										
3										
4										
5										
6										
7										
8										
9										
10										

Fill in the darkened squares.

X	1	2	3	4	5	6	7	8	9	10
1										
2										
3										
4										
5										
6										
7										
8										
9										
10										

Ringmaster Rudy's Review

Let's help Rudy fill in the blanks.
Remember 2 x 5 is the same as 5 x 2.
The order of the numbers does not matter when multiplying.

2 x <u>5</u> = 10	**2 x ___ = 16**	**10 x ___ = 90**
5 x ___ = 10	**8 x ___ = 16**	**9 x ___ = 90**
2 x ___ = 12	**2 x ___ = 8**	**10 x ___ = 10**
6 x ___ = 12	**4 x ___ = 8**	**1 x ___ = 10**
2 x ___ = 6	**2 x ___ = 18**	**10 x ___ = 50**
3 x ___ = 6	**9 x ___ = 18**	**5 x ___ = 50**
2 x ___ = 14	**2 x ___ = 2**	**10 x ___ = 20**
7 x ___ = 14	**1 x ___ = 2**	**2 x ___ = 20**
9 x ___ = 0	**2 x ___ = 4**	**7 x ___ = 0**

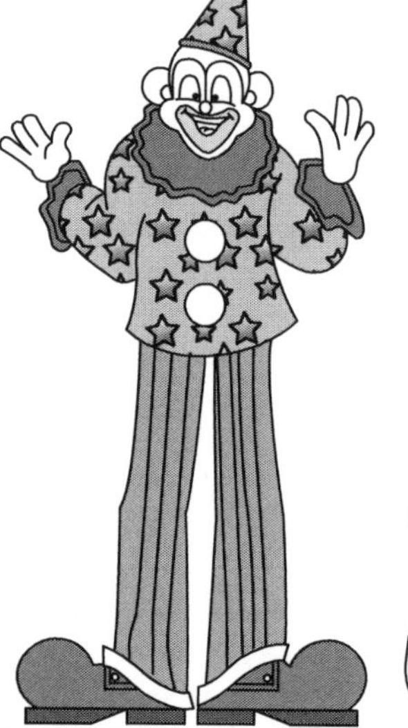

Help Rudy complete the pattern and find his way home to **100 Magic Circus Drive.**

Magic Circus Fun

Let's help Rudy with the Magic Circus.

How many clowns in total? **2** x **3** = **6**

How many giraffes in total? ____ x ____ = ____

How many bears in total? ____ x ____ = ____

How many monkeys in total? ____ x ____ = ____

If You Know the 2 x Table, You Know the 8's!

Fill in the 8 x table. Notice the last digits repeat the **2-4-6-8** pattern of the 2 x table in ***reverse*** order.
0 always follows the last number in the pattern.
The **8-6-4-2-0** pattern repeats after 40.
Fill in tables 2 and 8 across.

X	1	2	3	4	5	6	7	8	9	10
1		2						8		
2		4						16		
3		6						24		
4		8						32		
5		10						40		
6		12						4_		
7		14						5_		
8		16						6_		
9		18						7_		
10		20						8_		

Table 8 Review

Fill in the columns for tables 2 and 8.
Notice the multiples of the 8 x table
are 4 times larger than the 2's.
Example: 2 & 8, 10 & 40.
Why is that?

X	1	2	3	4	5	6	7	8	9	10
1										
2	2	4	6	8	10	12	14	16	18	20
3										
4										
5										
6										
7										
8	8	16	24	32	40	48	56	64	72	80
9										
10										

Magic Circus Fun

Help Gia complete the pattern and find her way home to **272 Magic Circus Drive.**

Clown Fun

Fill in the blanks.

8 x ___ = 24

3 x ___ = 24

2 x ___ = 18

9 x ___ = 18

8 x ___ = 56

7 x ___ = 56

8 x ___ = 40

5 x ___ = 40

2 x ___ = 14

7 x ___ = 14

8 x ___ = 32

4 x ___ = 32

2 x ___ = 12

6 x ___ = 12

8 x ___ = 48

6 x ___ = 48

8 x ___ = 16

2 x ___ = 16

8 x ___ = 72

9 x ___ = 72

Magic Circus Fun

Let's help Ringmaster Rudy
with the Magic Circus.

How many giraffes in total? ___ x ___ = ___

How many monkeys in total? ___ x ___ = ___

2 and 8 Review

Fill in the
darkened squares.

X	1	2	3	4	5	6	7	8	9	10
1										
2										
3										
4										
5										
6										
7										
8										
9										
10										

4-8-2-6-0
Now you know the 4 x table!

Fill in 4 x table. Notice the **4-8-2-6-0** pattern repeats after 20. Pretty cool, isn't it? Fill in 4 across.

X	1	2	3	4	5	6	7	8	9	10
1				4						
2				8						
3				12						
4				16						
5				20						
6				2_						
7				2_						
8				3_						
9				3_						
10				4_						

Grid Magic

Can you copy Rudy on the blank grid?

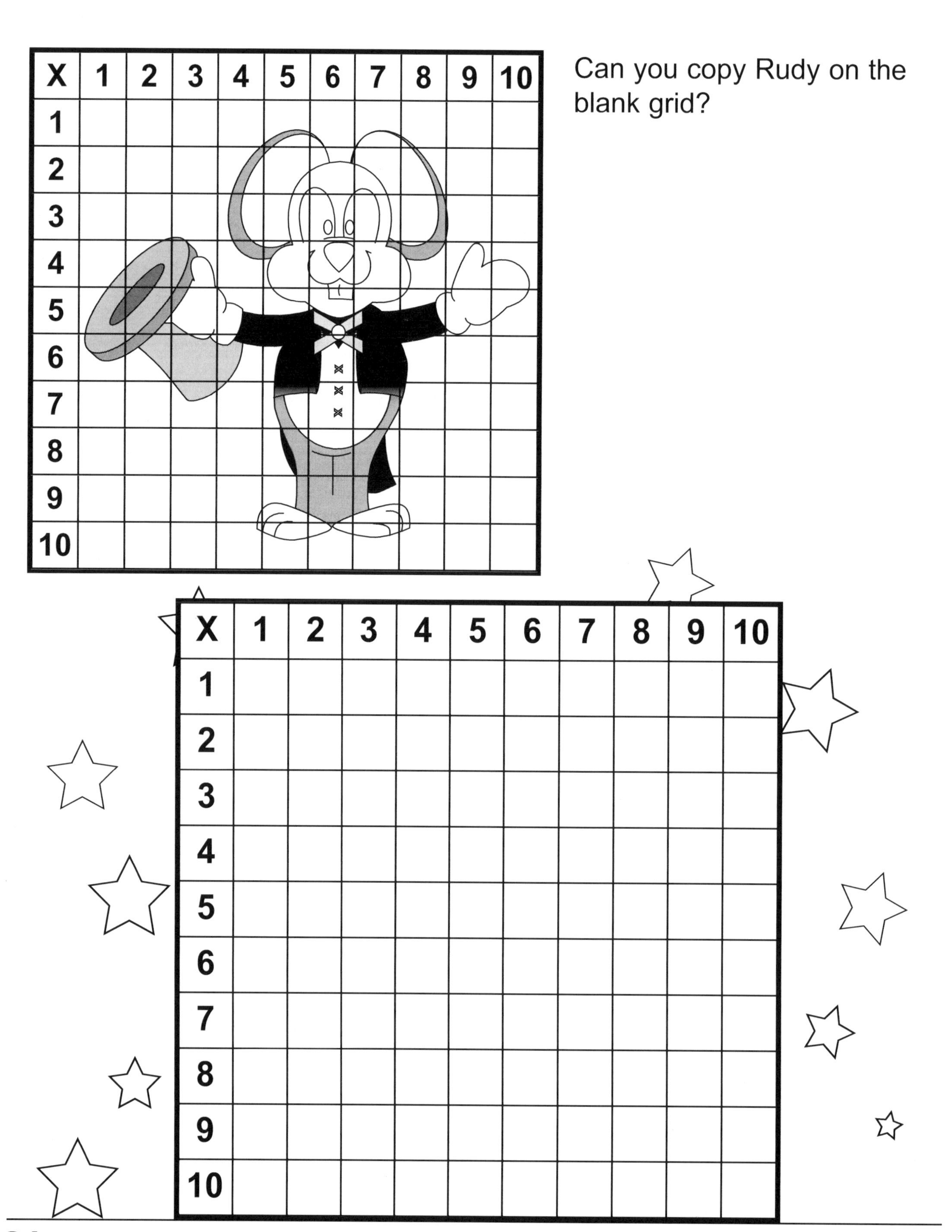

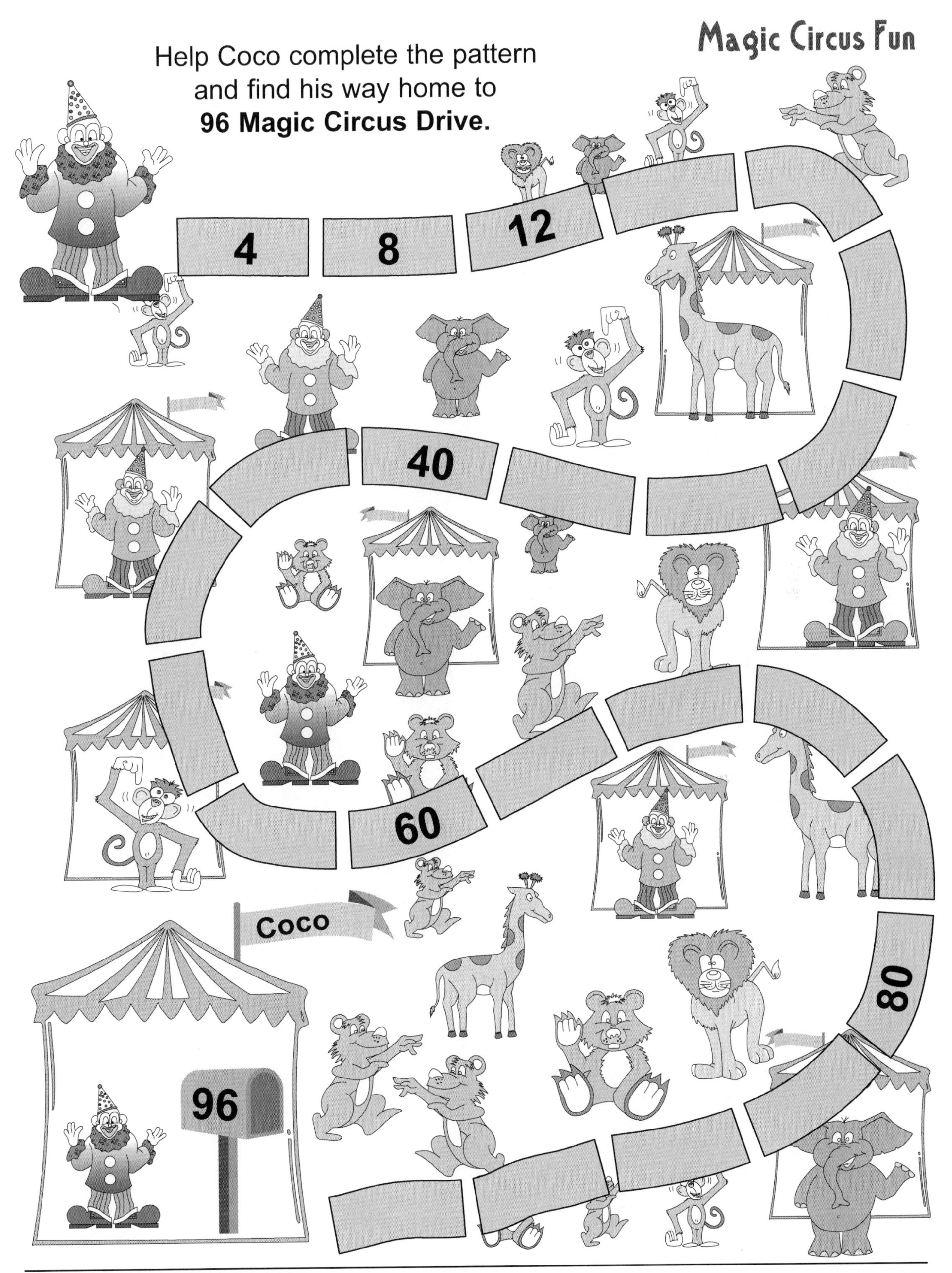
Magic Circus Fun
Help Coco complete the pattern
and find his way home to
96 Magic Circus Drive.
4
8
12
40
60
80
Coco
96

Clown Fun

Fill in the blanks.

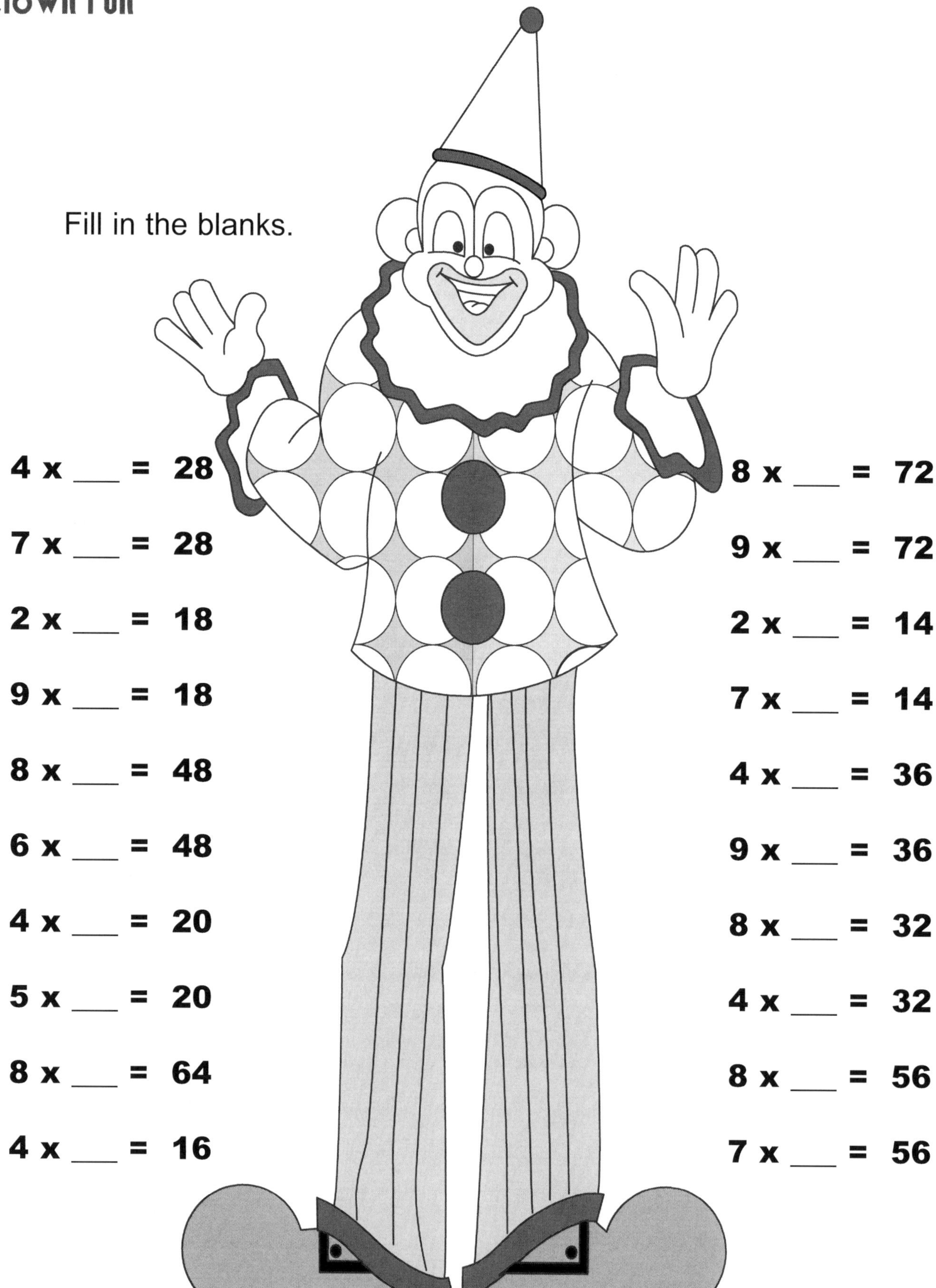

4 x ___ = 28

7 x ___ = 28

2 x ___ = 18

9 x ___ = 18

8 x ___ = 48

6 x ___ = 48

4 x ___ = 20

5 x ___ = 20

8 x ___ = 64

4 x ___ = 16

8 x ___ = 72

9 x ___ = 72

2 x ___ = 14

7 x ___ = 14

4 x ___ = 36

9 x ___ = 36

8 x ___ = 32

4 x ___ = 32

8 x ___ = 56

7 x ___ = 56

Circus Snacks

Hot Dogs $3	Ice Cream $2
Pretzels $1	Popcorn $3
Cotton Candy $2	Soda $1

Erica bought 8 pretzels at the circus for her friends. How much did she spend?

_____ x _____ = $_____

Tom bought his 4 brothers each a soda. How much did he spend?

_____ x _____ = $_____

The twins bought 4 bags of popcorn and 4 sodas. How much did they spend?

_____ x _____ = $_____

_____ x _____ = $_____

Total: _____ + _____ = $_____

Martha bought 8 ice cream bars and 10 sodas. How much did she spend?

_____ x _____ = $_____

_____ x _____ = $_____

Total: _____ + _____ = $_____

Carlos and Billy bought 4 hot dogs, 3 sodas and 2 ice creams. How much did they spend?

_____ x _____ = $_____

_____ x _____ = $_____

_____ x _____ = $_____

Total: ____+____+____ = $_____

Lucy bought her family 6 sodas, 5 pretzels and 8 hot dogs. How much did she spend?

_____ x _____ = $_____

_____ x _____ = $_____

_____ x _____ = $_____

Total: ____+____+____ = $_____

2 and 4 Magic!

Fill in the columns for 2 & 4. Notice the multiples of the 4 x table are 2 times larger than the 2's. Why is that?

X	1	2	3	4	5	6	7	8	9	10
1										
2	2	4	6	8	10	12	14	16	18	20
3										
4	4	8	12	16	20	24	28	32	36	40
5										
6										
7										
8										
9										
10										

4 and 8 Magic!

Fill in the columns for 4 & 8. Notice the multiples of the 8 x table are 2 times larger than the 4's. Why is that?

X	1	2	3	4	5	6	7	8	9	10
1										
2										
3										
4	4	8	12	16	20	24	28	32	36	40
5										
6										
7										
8	8	16	24	32	40	48	56	64	72	80
9										
10										

2 and 4 Magic Review

Fill in the darkened squares.

X	1	2	3	4	5	6	7	8	9	10
1										
2										
3										
4										
5										
6										
7										
8										
9										
10										

4 and 8 Magic Review

Fill in the darkened squares.

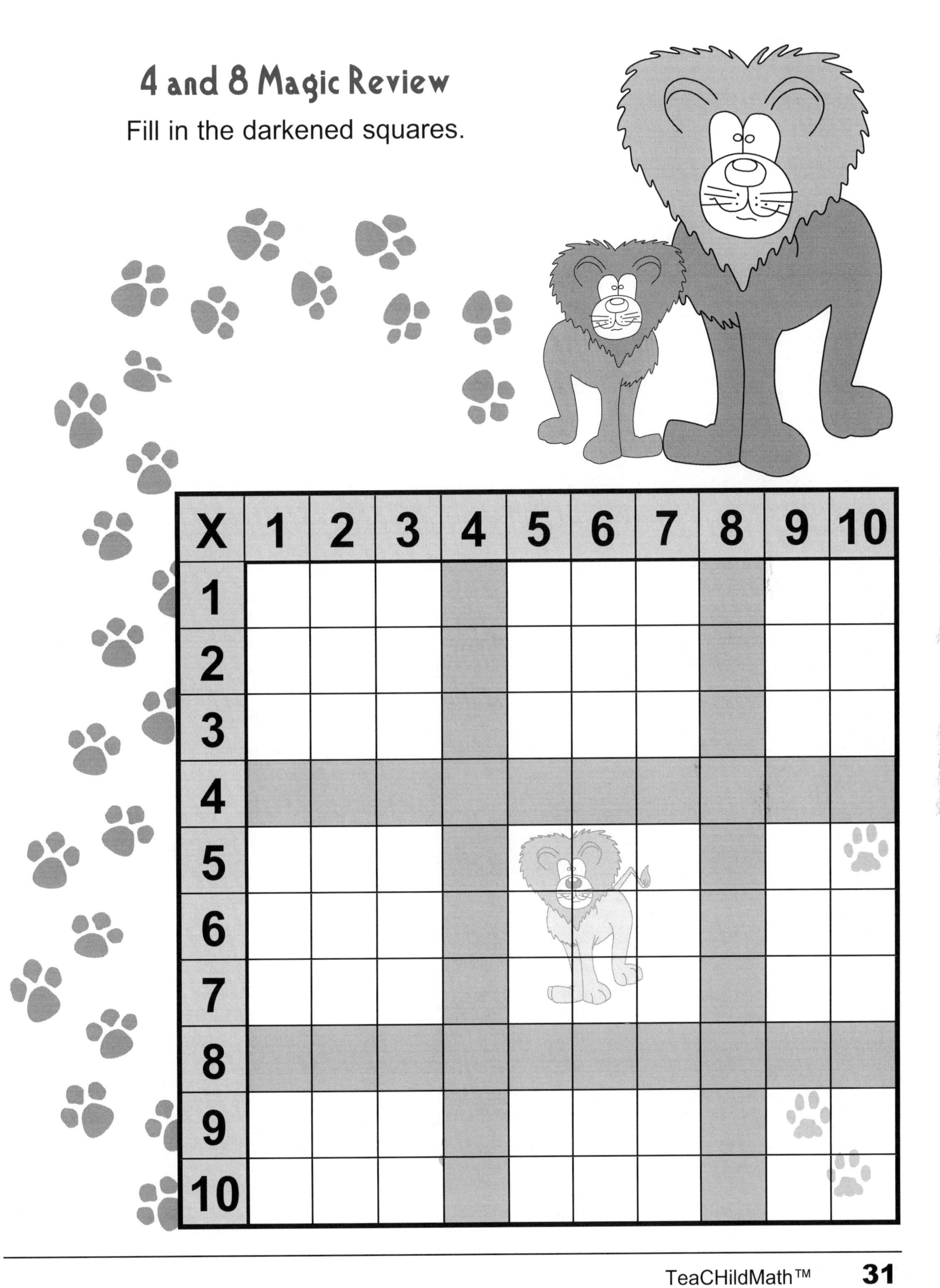

X	1	2	3	4	5	6	7	8	9	10
1										
2										
3										
4										
5										
6										
7										
8										
9										
10										

Circus Snack Survey

Children voted for their favorite Circus Snacks. Here are the results:

★ = 4 votes

1. Which snack got 28 votes? __________

2. How many votes did pretzels get? __________

3. Which snack got more votes than ice cream? __________

4. Which snack was the least favorite? __________

5. Which snacks got 36 votes? __________

6. Which was the most popular snack? __________

7. Which of the above is your favorite snack? __________

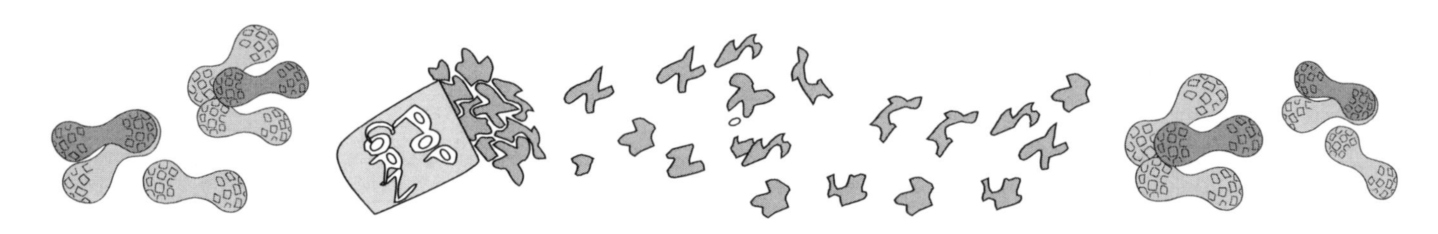

Color the Clown

Solve the problems to color the clown.

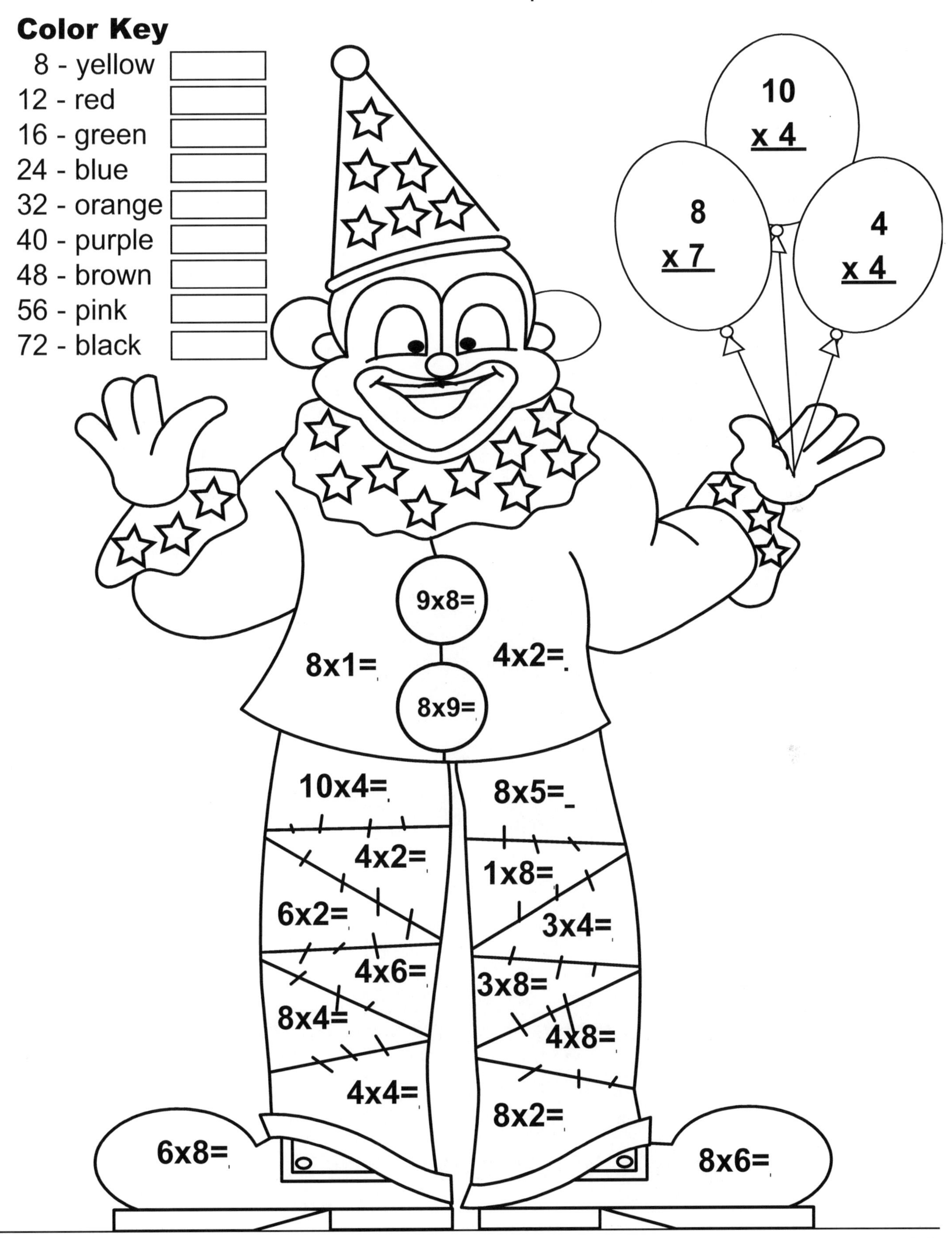

Clown Fun

Fill in the blanks.

2 x ___ = 16

4 x ___ = 28

8 x ___ = 40

2 x ___ = 14

8 x ___ = 48

4 x ___ = 16

8 x ___ = 72

4 x ___ = 20

8 x ___ = 24

2 x ___ = 20

8 x ___ = 32

2 x ___ = 18

4 x ___ = 36

8 x ___ = 56

2 x ___ = 10

4 x ___ = 12

8 x ___ = 16

4 x ___ = 24

8 x ___ = 64

4 x ___ = 40

Secret Code for Table 6?

Test your skill with tables 2, 4 and 8. Can you fill in 6?
It too has a pattern.

X		2		4		6		8
1		2		4		6		8
2		_		_		12		1_
3		_		1_		18		2_
4		_		1_		24		3_
5		10		20		30		40
6		1_		2_		3_		4_
7		1_		2_		4_		5_
8		1_		3_		4_		6_
9		1_		3_		5_		7_
10		20		40		60		80

Wouldn't you know? It's 6-2-8-4-0!

Fill in the 6 x table. Notice the 6-2-8-4-0 pattern repeats after 30. Fill in 6 across.

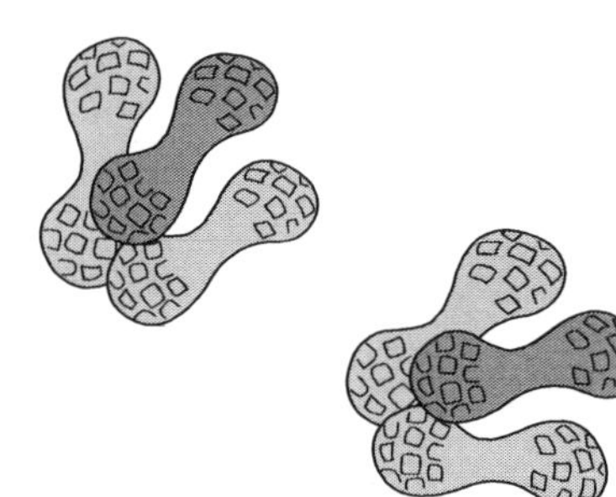

X	1	2	3	4	5	6	7	8	9	10
1						6				
2						12				
3						18				
4						24				
5						30				
6						3_				
7						4_				
8						4_				
9						5_				
10						6_				

PEANUTS

ELEPHANT BRAND

Help Tiny complete the pattern and find her way home to **150 Magic Circus Drive.**

Circus Snack Survey

Children voted for their favorite Circus Snacks. Here are the results:

★ = 6 votes

1. Which snack got 36 votes? _________

2. How many votes did ice cream get? _________

3. Which snack got more votes than ice cream? _________

4. Which snack was the least favorite? _________

5. Which snack got 42 votes? _________

6. Which was the most popular snack? _________

7. Which of the above is your favorite snack? _________

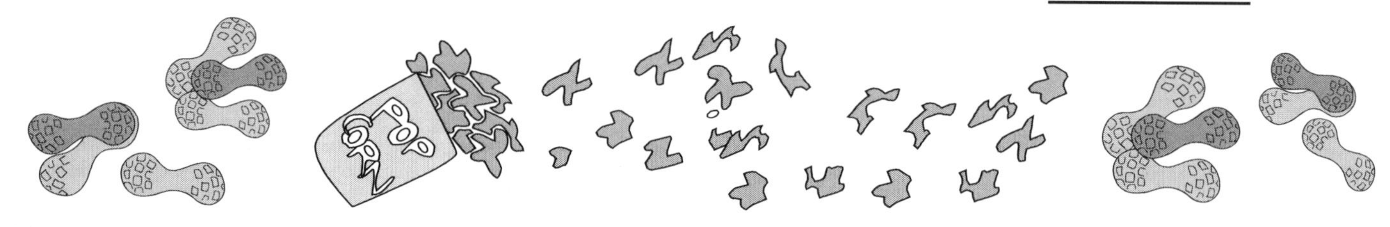

2, 4, 6 and 8 Magic!

Fill in the columns for 2, 4, 6 & 8.

X	1	2	3	4	5	6	7	8	9	10
1										
2	2	4	6	8	10	12	14	16	18	20
3										
4	4	8	12	16	20	24	28	32	36	40
5										
6	6	12	18	24	30	36	42	48	54	60
7										
8	8	16	24	32	40	48	56	64	72	80
9										
10										

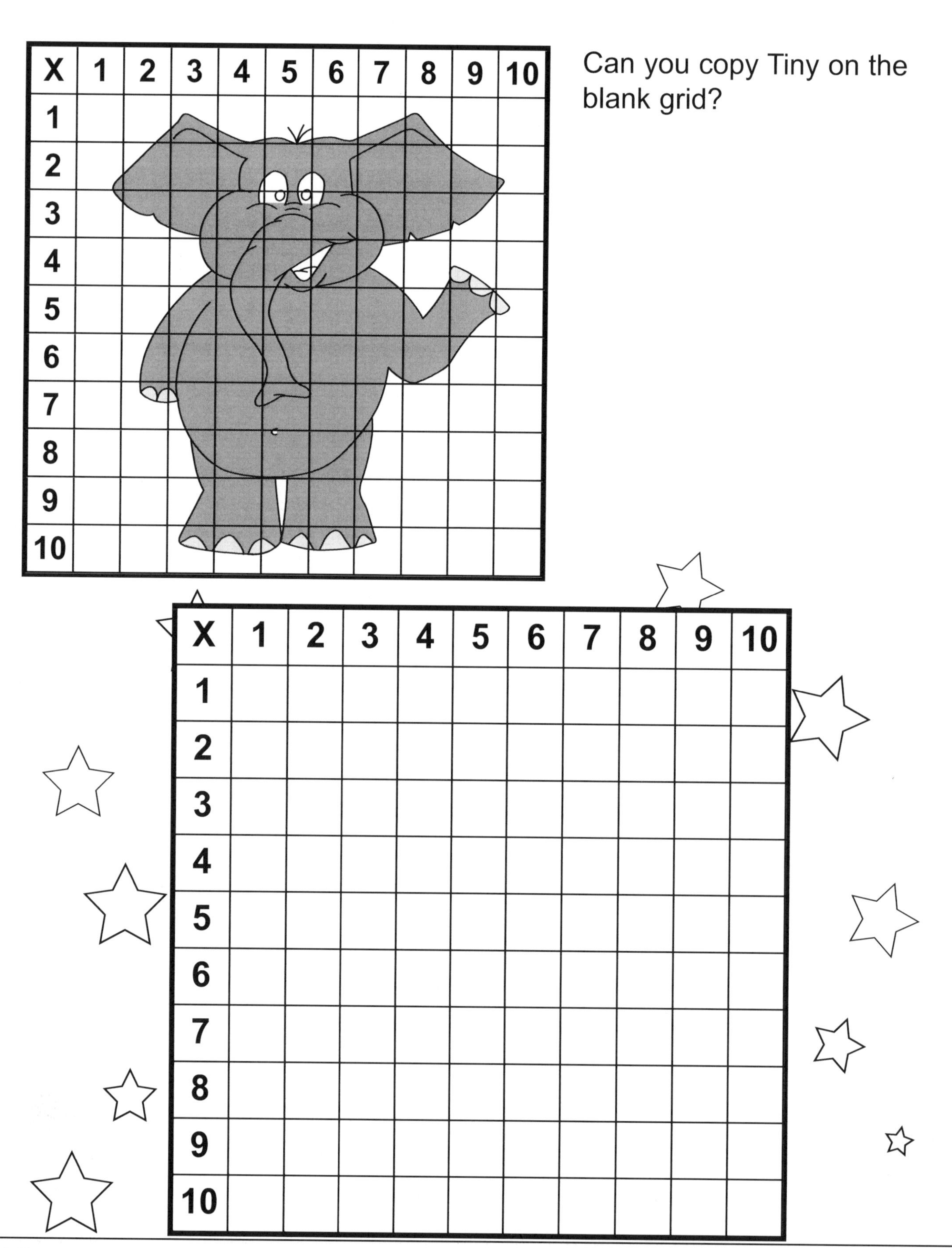

Can you copy Tiny on the blank grid?

The pattern for table 8 is the ***reverse*** of table 2, followed by 0.
The pattern for table 6 is the ***reverse*** of table 4, followed by 0.
Can you fill in the rest?

X	2		8
1	2		8
2	4		_6
3	6		_4
4	8		_2
5	_0		_0
6	_2		_8
7	_4		_6
8	_6		_4
9	_8		_2
10	_0		_0

4		6
4		6
8		_2
_2		_8
_6		_4
_0		_0
_4		_6
_8		_2
_2		_8
_6		_4
_0		_0

Pattern Review

Remember the patterns?

2 x table : 2-4-6-8 followed by **0**.
8 x table : 8-6-4-2 followed by **0**.
4 x table : 4-8-2-6 followed by **0**.
6 x table : 6-2-8-4 followed by **0**.

Can you fill in the rest?

X	2		8
1	2		8
2	4		16
3	6		24
4	8		32
5	10		40
6	1_		4_
7	1_		5_
8	1_		6_
9	1_		7_
10	2_		8_

4		6
4		6
8		12
12		18
16		24
20		30
2_		3_
2_		4_
3_		4_
3_		5_
4_		6_

There are ____ elephants in the ring. Each elephant has ____ balloons. How many balloons are there?

_____ x _____ = _____

There are ____ clowns in the ring. Each clown has ____ balls. How many balls are there?

_____ x _____ = _____

There are ____ monkeys in the ring. Each monkey has ____ lollipops. How many lollipops are there?

_____ x _____ = _____

Can you copy Sam on the blank grid?

Rudy's Secret Code for 2, 4, 6 and 8!

1st secret clue: the pattern repeats after 10, 20, 30 and 40.
2nd secret clue: 8 x table pattern is 2 x table in **reverse**.
3rd secret clue: 6 x table pattern is 4 x table in **reverse**.
4th secret clue: tables 2, 4, 6 & 8 end in some combination of **2-4-6-8** followed by **0**.

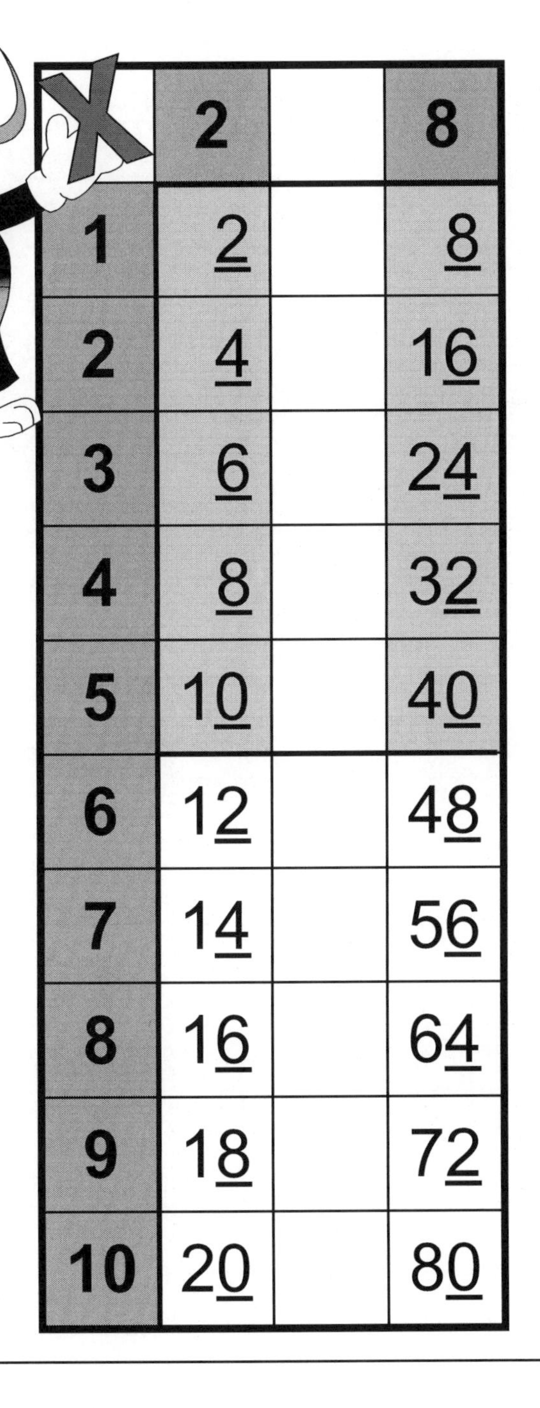

X	2		8
1	2		8
2	4		16
3	6		24
4	8		32
5	10		40
6	12		48
7	14		56
8	16		64
9	18		72
10	20		80

4		6
4		6
8		12
12		18
16		24
20		30
24		36
28		42
32		48
36		54
40		60

Remember the Secret Code?

Secret Clue: patterns repeat after 10, 20, 30 & 40.

X	2		8
1	2		8
2	_		1_
3	_		2_
4	_		3_
5	1_		4_
6	1_		4_
7	1_		5_
8	1_		6_
9	1_		7_
10	2_		8_

4		6
4		6
_		1_
1_		1_
1_		2_
2_		3_
2_		3_
2_		4_
3_		4_
3_		5_
4_		6_

Color the Clown

Solve the problems to color the clown.

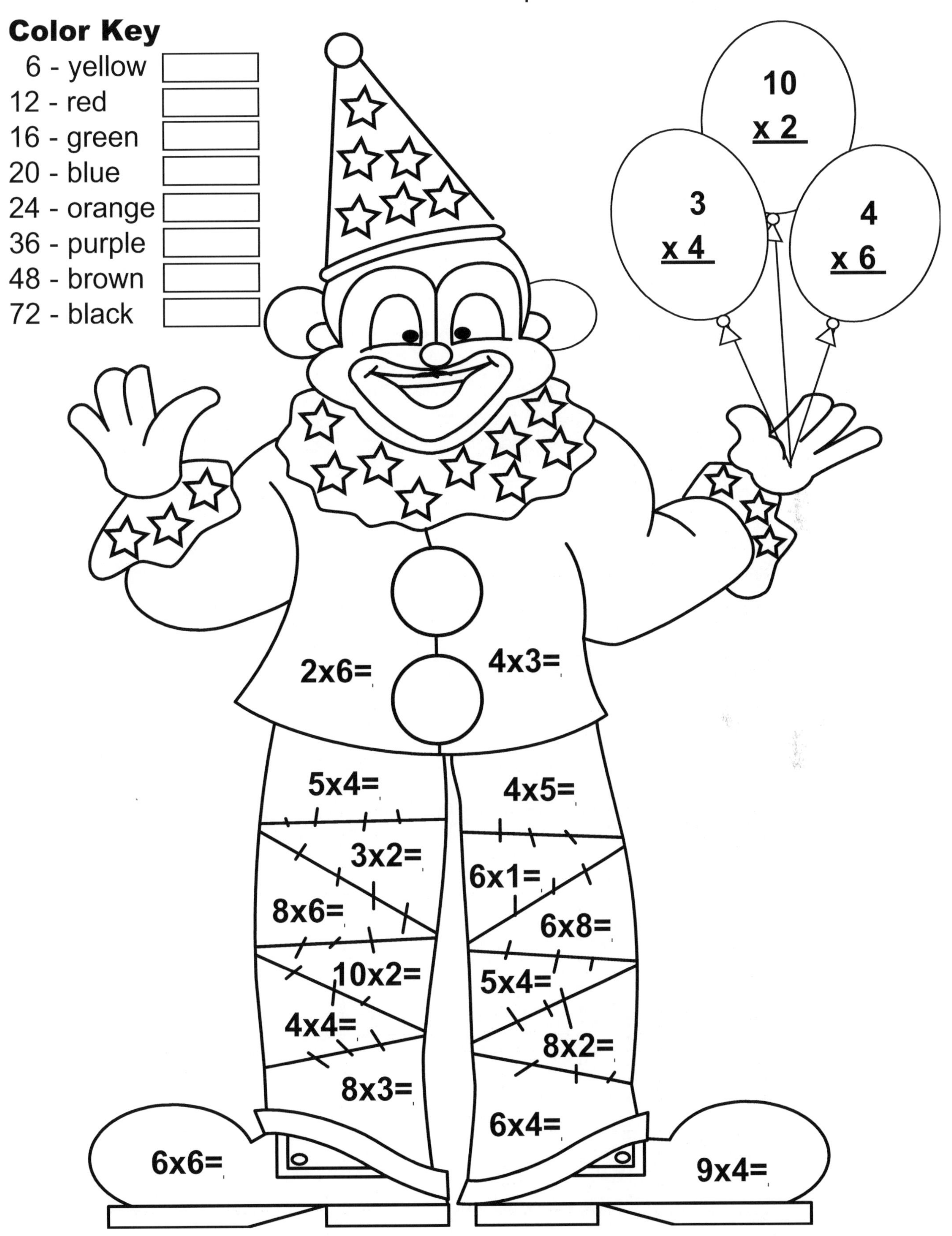

Secret Code Challenge!

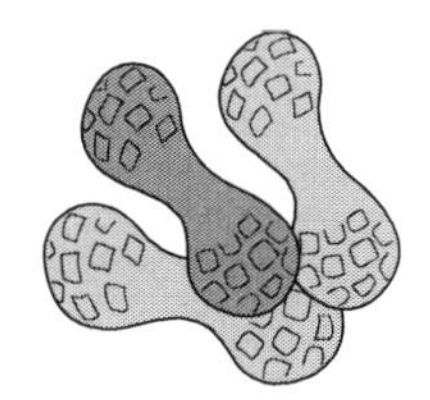

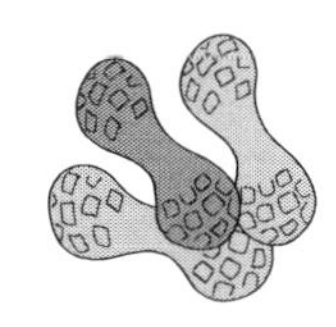

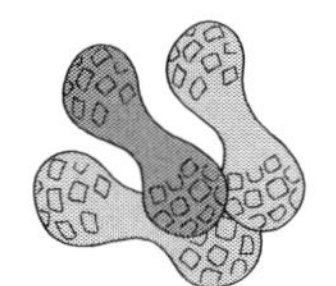

Fill in the columns.

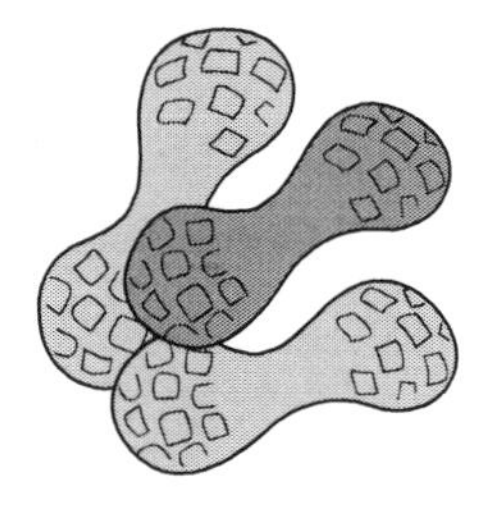

X	2		8
1	2		8
2	_		_
3	_		_
4	_		_
5	10		40
6	_		_
7	_		_
8	_		_
9	_		_
10	20		80

4		6
4		6
_		_
_		_
_		_
20		30
_		_
_		_
_		_
_		_
40		60

Let's help Rudy solve the following:

8 x 3 =___	**4 x ___ = 24**	**8 x 7 = ___**
2 x 7 = ___	**4 x ___ = 36**	**9 x 6 = ___**
4 x 4 = ___	**5 x ___ = 30**	**6 x 7 = ___**
2 x 8 = ___	**8 x ___ = 32**	**9 x 8 = ___**
0 x 8 = ___	**8 x ___ = 72**	**6 x 8 = ___**
2 x 3 = ___	**4 x ___ = 40**	**8 x 8 = ___**
8 x 5 = ___	**6 x ___ = 24**	**8 x 6 = ___**
4 x 3 = ___	**3 x ___ = 18**	**6 x 3 = ___**
2 x ___ = 12	**4 x ___ = 32**	**8 x ___ = 56**
5 x ___ = 10	**5 x ___ = 20**	**6 x ___ = 36**
6 x ___ = 12	**8 x ___ = 16**	**2 x ___ = 18**
2 x ___ = 20	**3 x ___ = 24**	**6 x ___ = 30**
9 x ___ = 36	**5 x ___ = 40**	**10 x ___ = 100**

2, 4, 6 and 8 Magic Review

Fill in the darkened squares.

Even Number Search

Circle all the EVEN multiples.
What did you discover?

EVEN number x **EVEN** number = _____
EVEN number x **ODD** number = _____

X	1	2	3	4	5	6	7	8	9	10
1	1	2	3	4	5	6	7	8	9	10
2	2	4	6	8	10	12	14	16	18	20
3	3	6	9	12	15	18	21	24	27	30
4	4	8	12	16	20	24	28	32	36	40
5	5	10	15	20	25	30	35	40	45	50
6	6	12	18	24	30	36	42	48	54	60
7	7	14	21	28	35	42	49	56	63	70
8	8	16	24	32	40	48	56	64	72	80
9	9	18	27	36	45	54	63	72	81	90
10	10	20	30	40	50	60	70	80	90	100

Decoding Patterns?

Write the multiplication problem for:

2 rows of 10 s = ____ s

2 x 10 = ____

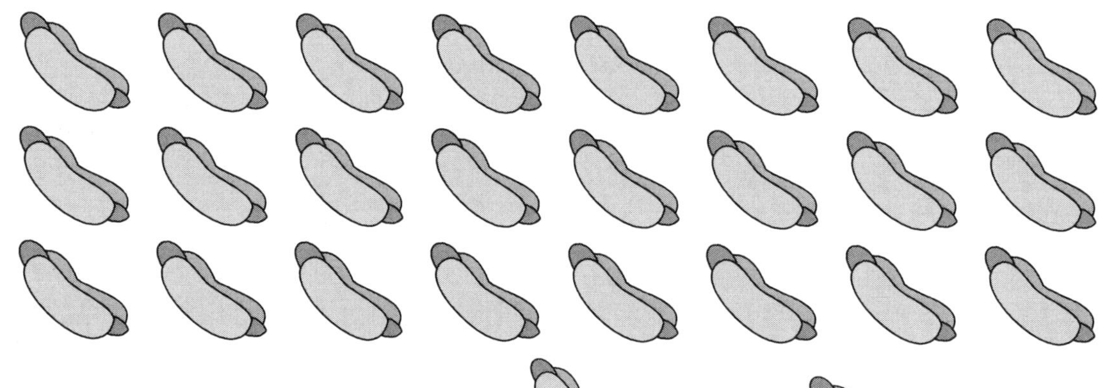

___ rows of ___ s = ____ s

___ x ___ = ____

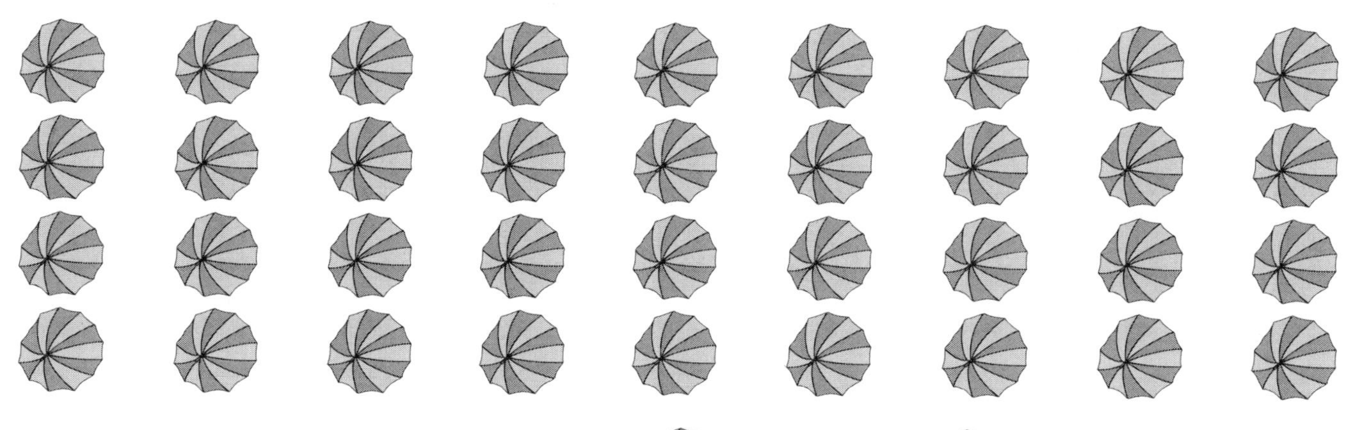

___ rows of ___ s = ____ s

___ x ___ = ____

Fill in tables 2, 4, 6, 8 and 10.

X	1	2	3	4	5	6	7	8	9	10
1										
2										
3										
4										
5										
6										
7										
8										
9										
10										

Race to the Finish

Fill in the blanks.

3 x ___ = 24

4 x ___ = 12

1 x ___ = 10

6 x ___ = 18

2 x ___ = 16

7 x ___ = 14

5 x ___ = 40

8 x ___ = 72

4 x ___ = 24

5 x ___ = 50

6 x ___ = 24

4 x ___ = 36

8 x ___ = 56

5 x ___ = 30

9 x ___ = 72

4 x ___ = 16

8 x ___ = 40

9 x ___ = 36

3 x ___ = 12

8 x ___ = 32

8 x ___ = 48

2 x ___ = 18

8 x ___ = 64

Good Job!

6 x ___ = 42

7 x ___ = 28

2 x ___ = 10

3 x ___ = 30

5 x ___ = 10

6 x ___ = 54

4 x ___ = 20

6 x ___ = 48

3 x ___ = 18

2 x ___ = 20

6 x ___ = 36

2 x ___ = 14

Odd or Even?

Even X Even = **EVEN**

Even X Odd = **EVEN** Odd X Even = **EVEN**

Odd X Odd = **ODD**

Notice the **hopscotch pattern** of the **odd** multiples. Turn the page sideways. Notice 4 x 2 is the **same** as 2 x 4. Check out the other dot multiples.

X	1	2	3	4	5
1	•	••	•••	••••	•••••
2	• •	•• ••	••• •••	•••• ••••	••••• •••••
3	• • •	•• •• ••	••• ••• •••	•••• •••• ••••	••••• ••••• •••••
4	• • • •	•• •• •• ••	••• ••• ••• •••	•••• •••• •••• ••••	••••• ••••• ••••• •••••
5	• • • • •	•• •• •• •• ••	••• ••• ••• ••• •••	•••• •••• •••• •••• ••••	••••• ••••• ••••• ••••• •••••

Can You Multiply the DOTS?

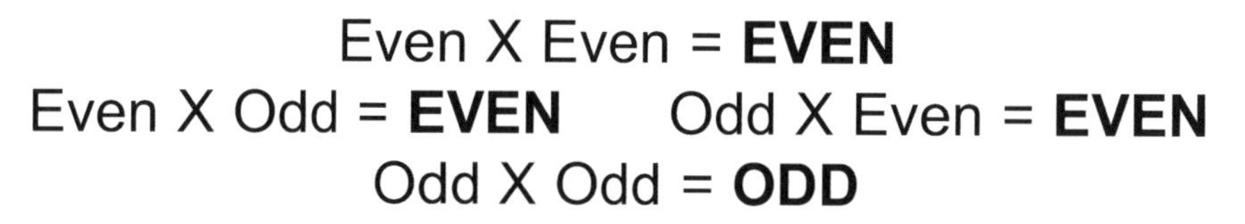

Even X Even = **EVEN**

Even X Odd = **EVEN**   Odd X Even = **EVEN**

Odd X Odd = **ODD**

Fill in with the correct dot multiple. Notice the DARK squares represent ODD multiples.

X	1	2	3	4	5
1	•	••	•••	••••	•••••
2					
3					
4					
5					

Help Rex complete the pattern and find his way home to **99 Magic Circus Drive.**

Odd or Even?

Circle each pair of 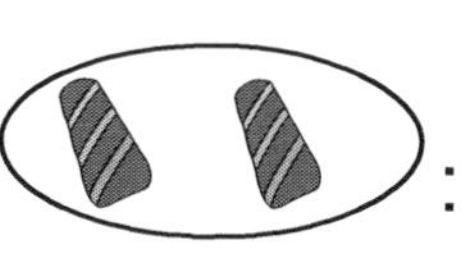:

8 bikes 4 times = 32 bikes

8 X 4 = 32

When you grouped in pairs, was any bike left over?

Yes ____ No ✓

32 is an even number.

Do the same with :

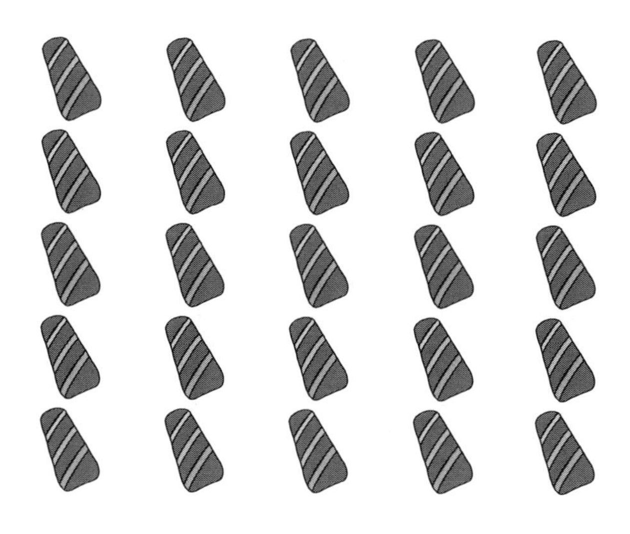

5 s 5 times = ____ s

5 X 5 = 25

When you grouped in 2's, was any left over?

Yes ____ No ___

25 is an ______ number.

With :

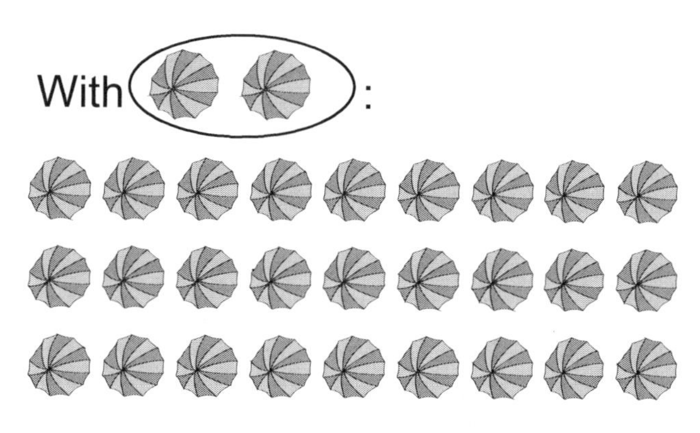

9 s 3 times = ____ s

9 X 3 = 27

Was any left over?

Yes ____ No____

Odd ____ or Even ____

1, 3, 5, 7, 9 or any number ending in 1, 3, 5, 7, 9 are **ODD**.
2, 4, 6, 8, 0 or any number ending in 2, 4, 6, 8, 0 are **EVEN**.

Circle each pair of  :

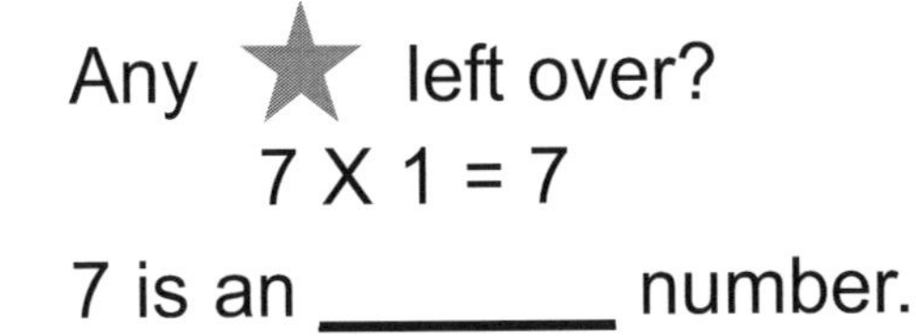

Any ★ left over?

7 X 1 = 7

7 is an _______ number.

Any ★ left over?

7 X 2 = 14

14 is an _______ number.

Any ★ left over?

7 X 3 = 21

21 is an _______ number.

Secret Clues to Multiplication:

Odd number x **Odd = ODD**

Odd number x **Even = EVEN**

Even number x **Even = EVEN**

Even Number x **ANY** number = **EVEN**

Odd x Odd = ODD

All other multiples are Even:

Odd x Even = EVEN

Even x Odd = EVEN

Even x Even = EVEN

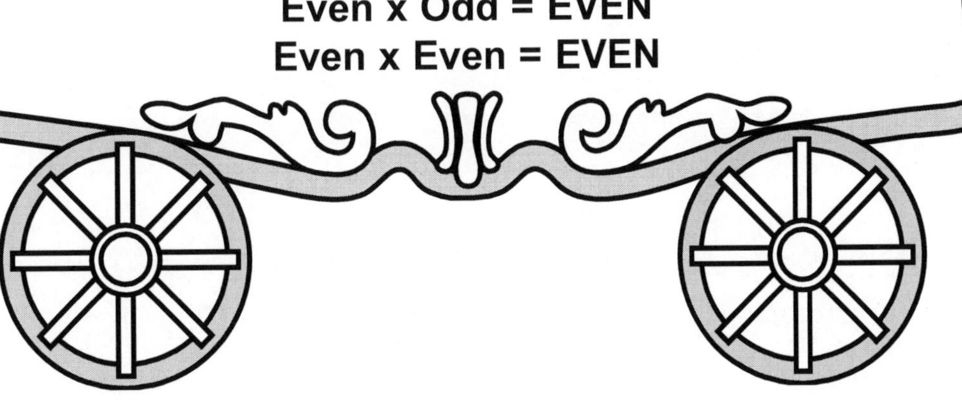

Odd or Even Multiple?

Secret Clue: If one number is **even**, the multiple is **even**.
Circle every **even** number first. Fill in with **odd** or **even**.

[6] x 3 = even	3 x 7 = odd	[8] x [2] = even
7 x 7 = ____	9 x 8 = ____	5 x 3 = ____
9 x 3 = ____	8 x 8 = ____	4 x 7 = ____
7 x 9 = ____	2 x 9 = ____	5 x 8 = ____
5 x 6 = ____	9 x 9 = ____	6 x 4 = ____
7 x 5 = ____	2 x 6 = ____	9 x 1 = ____
6 x 7 = ____	5 x 5 = ____	3 x 3 = ____
3 x 2 = ____	9 x 4 = ____	6 x 9 = ____
4 x 1 = ____	7 x 8 = ____	8 x 3 = ____
10 x 7 = ____	1 x 8 = ____	4 x 3 = ____
6 x 8 = ____	2 x 7 = ____	9 x 5 = ____
2 x 2 = ____	4 x 5 = ____	7 x 1 = ____

Multiplying Odds and Evens

Fill in each square with an **e** for **even** or an **o** for **odd**.
Notice the **hopscotch** pattern for **ODD** numbers.

Amazing fact:
Even x Even = Even
Even x Odd = Even
Odd x Even = Even
Odd x Odd = Odd

X	1	2	3	4	5	6	7	8	9	10
1	o	e								
2	e	e								
3										
4										
5										
6										
7										
8										
9										
10										

Table 5 Hopscotch 5-0-5-0!

Fill in the 5 column. **Clue:** last digits have a **5 - 0 pattern.**
Remember: **5** x an **EVEN** number ends in **0.**
5 x an **ODD** number ends in **5.**
Fill in across. Check your answers with the 5 column.

X	1	2	3	4	5	6	7	8	9	10
1					5					
2					1_					
3					15					
4					2_					
5					25					
6					3_					
7					35					
8					4_					
9					45					
10					5_					

Help Sam complete the pattern
and find his way home to
155 Magic Circus Drive.

Circus Fun

Help Rudy multiply the following.
Underneath, write the ODD/EVEN rule.

4 x 4 = ____ e x e = EVEN	5 x 5 = ____ ___ x ___ = ____	6 x ___ = 48 ___ x ___ = ____
6 x ___ = 12 ___ x ___ = ____	8 x 3 = ____ ___ x ___ = ____	5 x ___ = 30 ___ x ___ = ____
6 x 7 = ____ ___ x ___ = ____	5 x 10 = ____ ___ x ___ = ____	6 x ___ = 54 ___ x ___ = ____
8 x ___ = 40 ___ x ___ = ____	8 x ___ = 80 ___ x ___ = ____	5 x ___ = 15 ___ x ___ = ____
4 x ___ = 32 ___ x ___ = ____	6 x ___ = 24 ___ x ___ = ____	5 x ___ = 45 ___ x ___ = ____
5 x 7 = ____ ___ x ___ = ____	8 x ___ = 64 ___ x ___ = ____	4 x 9 = ____ ___ x ___ = ____
8 x ___ = 72 ___ x ___ = ____	6 x ___ = 36 ___ x ___ = ____	5 x 4 = ____ ___ x ___ = ____

5 and 10 Magic!

Fill in tables 5 and 10.
How much larger are the multiples of the 10 x table?
Example: 5 & 10, 30 & 60. Why is that?

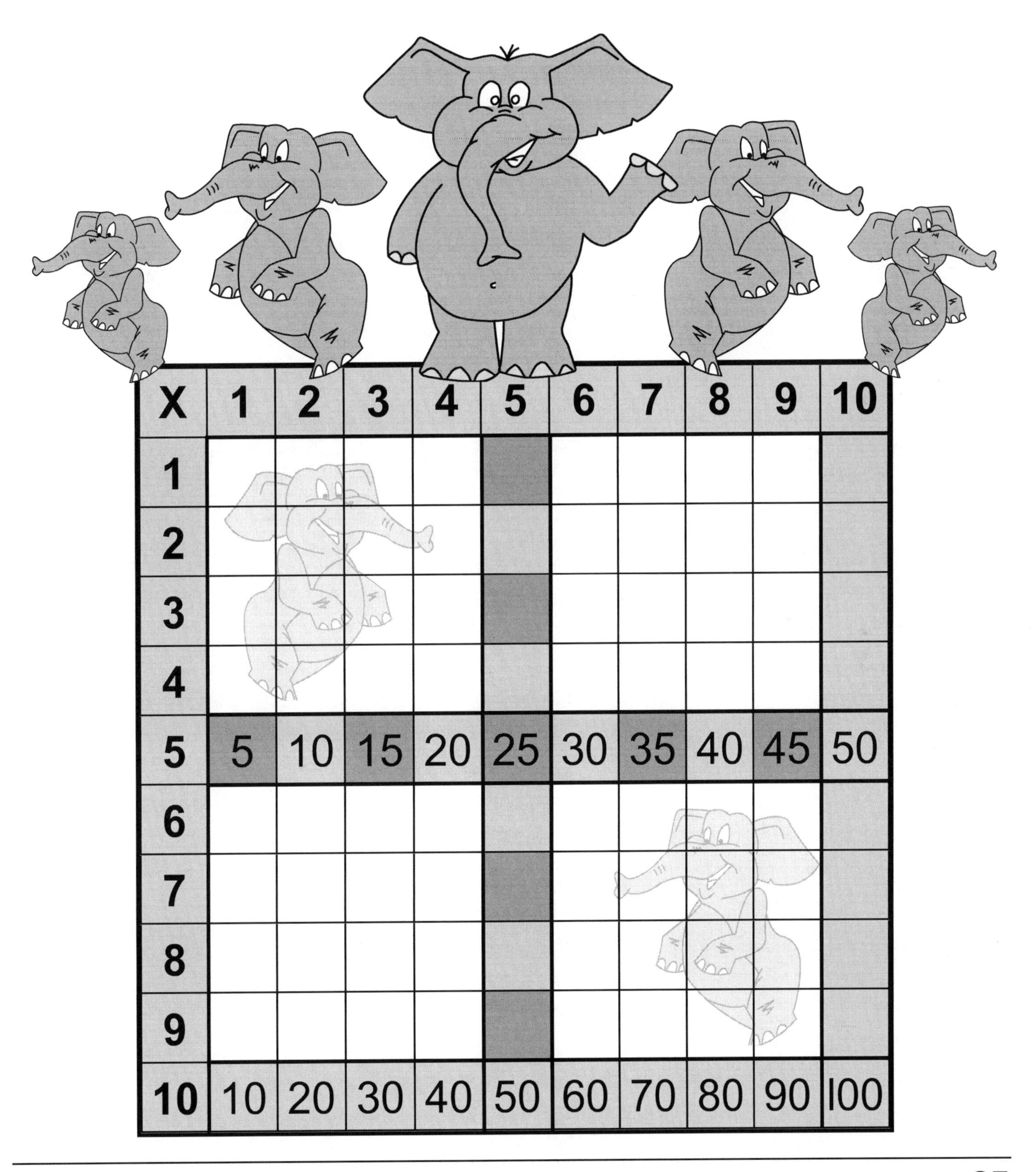

X	1	2	3	4	5	6	7	8	9	10
1										
2										
3										
4										
5	5	10	15	20	25	30	35	40	45	50
6										
7										
8										
9										
10	10	20	30	40	50	60	70	80	90	100

1, 5 and 10 Challenge!

Fill in tables 1, 5 and 10.

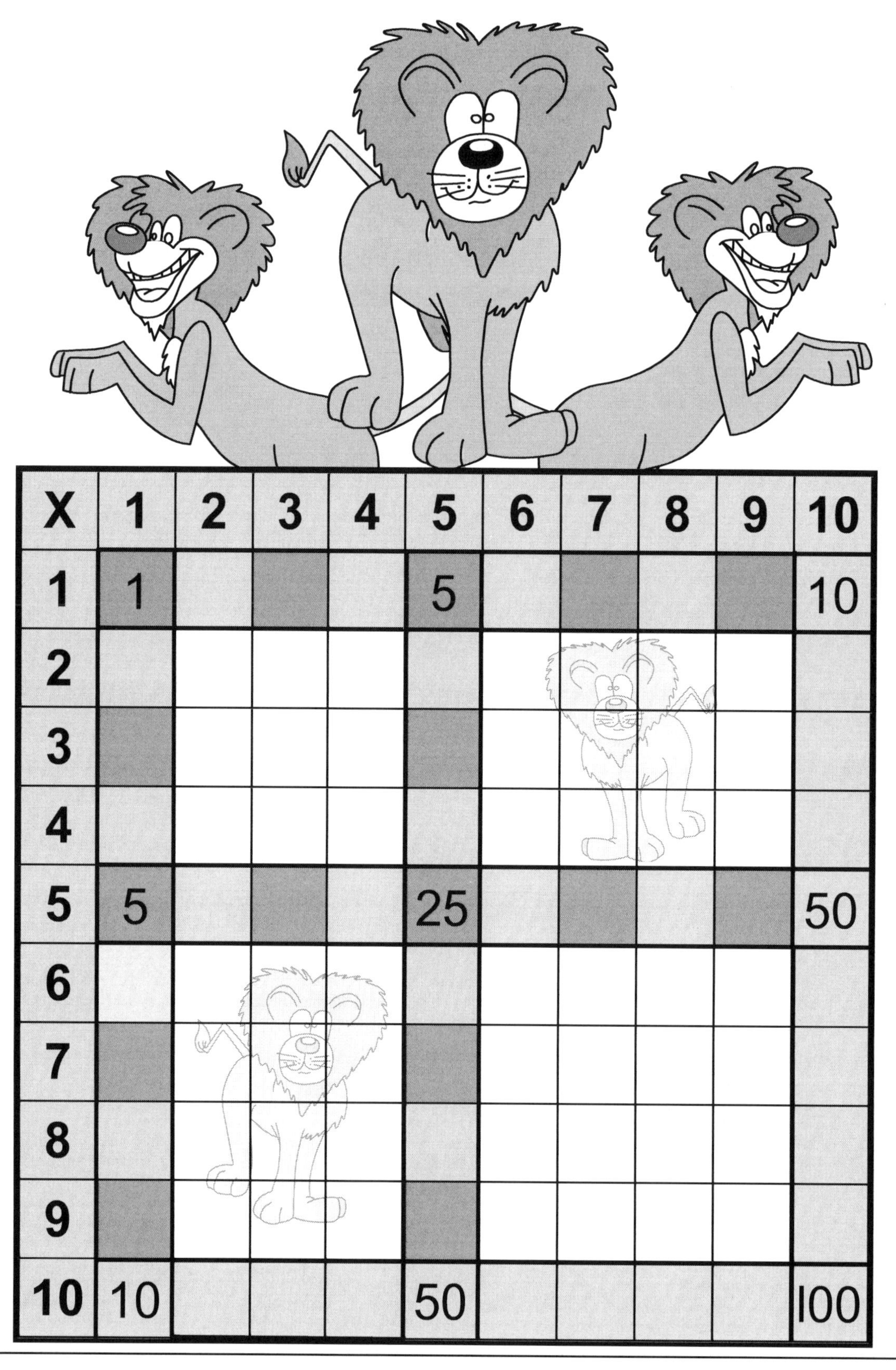

X	1	2	3	4	5	6	7	8	9	10
1	1				5					10
2										
3										
4										
5	5				25					50
6										
7										
8										
9										
10	10				50					100

Circus Snack Survey

Children voted for their favorite Circus Snacks. Here are the results:

★ = 5 votes

1. Which snack got 45 votes? _________

2. How many votes did peanuts get? _________

3. Which snack got fewer votes than pretzels? _________

4. Which snack was the least favorite? _________

5. Which snack got 40 votes? _________

6. Which was the most popular snack? _________

7. Which of the above is your favorite snack? _________

Here's the Answer!

What's the Problem?

Fill in the blanks. If an answer appears twice, give another possibility using different numbers. Do not multiply by 1.

Example: **<u>3 x 4</u>** = 12 or **<u>6 x 2</u>** = 12

___ x ___ = 25	**___ x ___ = 20**	**___ x ___ = 20**
___ x ___ = 24	**___ x ___ = 24**	**___ x ___ = 72**
___ x ___ = 40	**___ x ___ = 40**	**___ x ___ = 48**
___ x ___ = 30	**___ x ___ = 30**	**___ x ___ = 36**
___ x ___ = 16	**___ x ___ = 16**	**___ x ___ = 60**
___ x ___ = 12	**___ x ___ = 12**	**___ x ___ = 56**
___ x ___ = 54	**___ x ___ = 42**	**___ x ___ = 28**
___ x ___ = 35	**___ x ___ = 50**	**___ x ___ = 80**
___ x ___ = 64	**___ x ___ = 15**	**___ x ___ = 45**
___ x ___ = 18	**___ x ___ = 18**	**___ x ___ = 8**
___ x ___ = 10	**___ x ___ = 6**	**___ x ___ = 4**
___ x ___ = 32	**___ x ___ = 0**	**___ x ___ = 70**
___ x ___ = 80	**Good Job!**	**___ x ___ = 14**

Color the Clown

Solve the problems to color the clown.

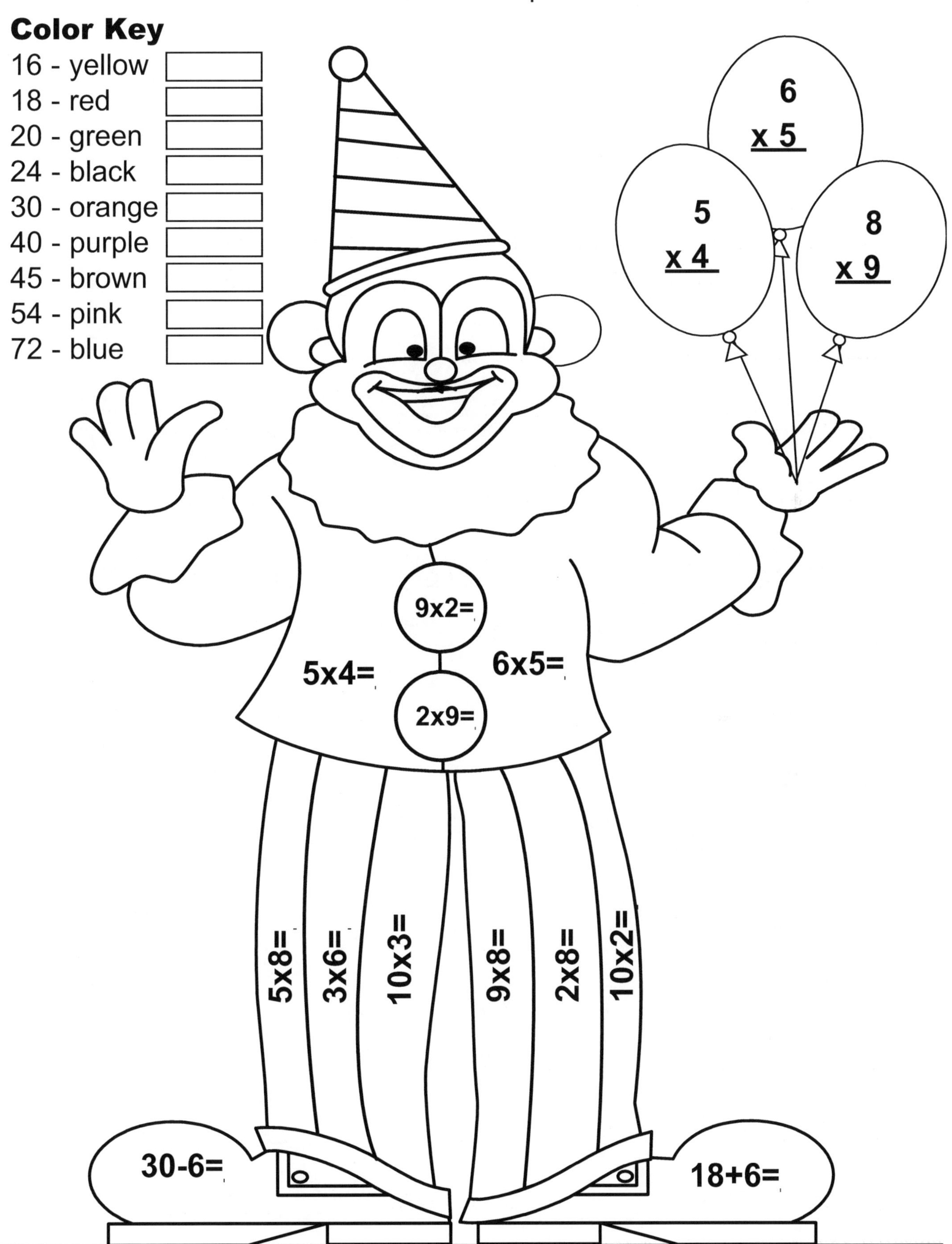

Help Rudy solve the following:

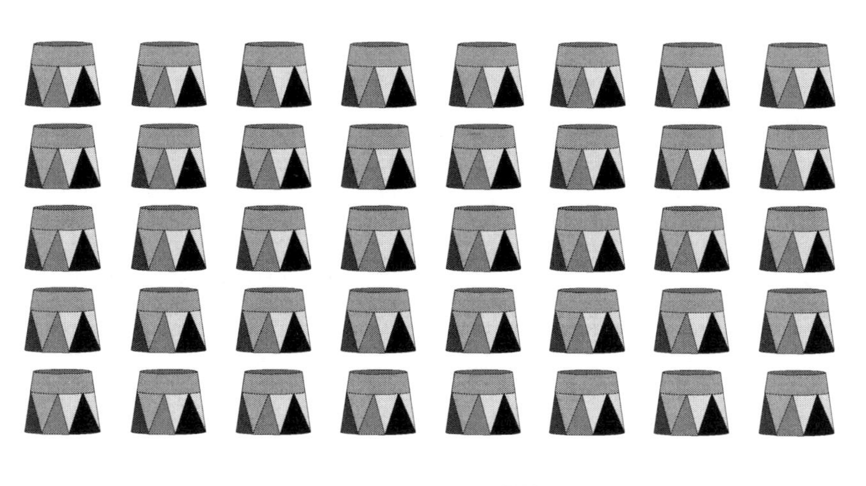

__ x __ = __

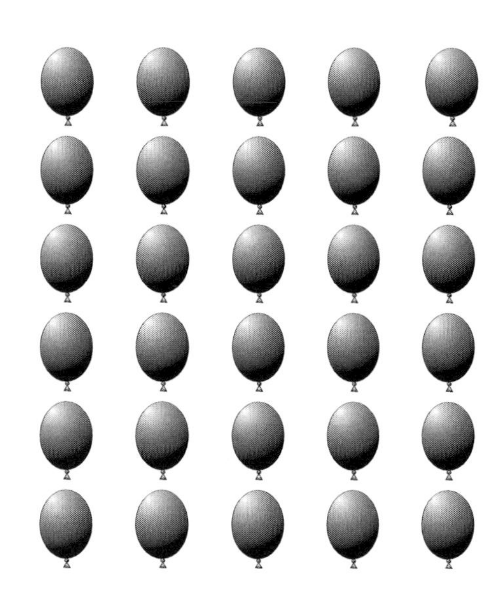

__ x __ = __

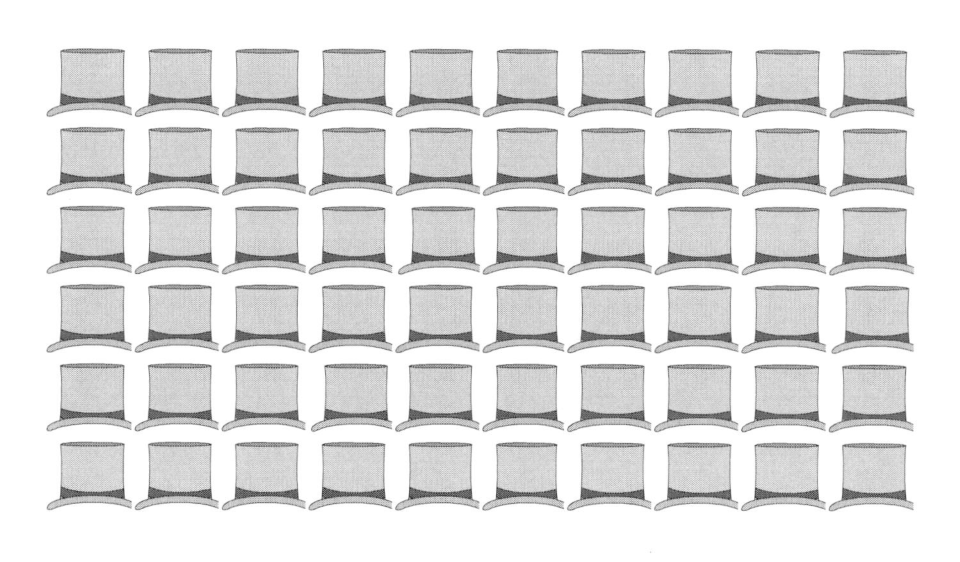

__ x __ = __

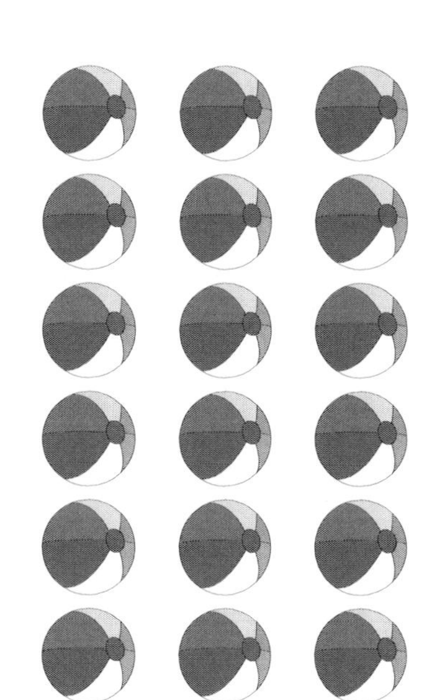

__ x __ = __

Circus Fun

Help Rudy multiply the following.
Underneath, write the Odd/Even rule.

5 x 5 = _____	10 x 3 = _____	7 x ___= 35
O x O = ODD	___ x ___= _____	___ x ___ = _____
6 x ___ = 30	2 x ___ = 20	8 x 5 = _____
___ x ___ = _____	___ x ___ = _____	___ x ___ = _____
3 x 5 = _____	10 x 10 = _____	2 x ___= 10
___ x ___= _____	___ x ___= _____	___ x ___ = _____
4 x 5 = _____	5 x ___ = 50	9 x ___ = 45
___ x ___= _____	___x ___ = _____	___ x ___ = _____
5 x ___= 40	5 x ___ = 25	6 x ___ = 60
___ x ___= _____	___x ___ = _____	___x ___ = _____

Multiply:

5	5	5	5	8	4	7	3	10
x2	x5	x6	x9	x5	x5	x5	x5	x5

Monkey Fun

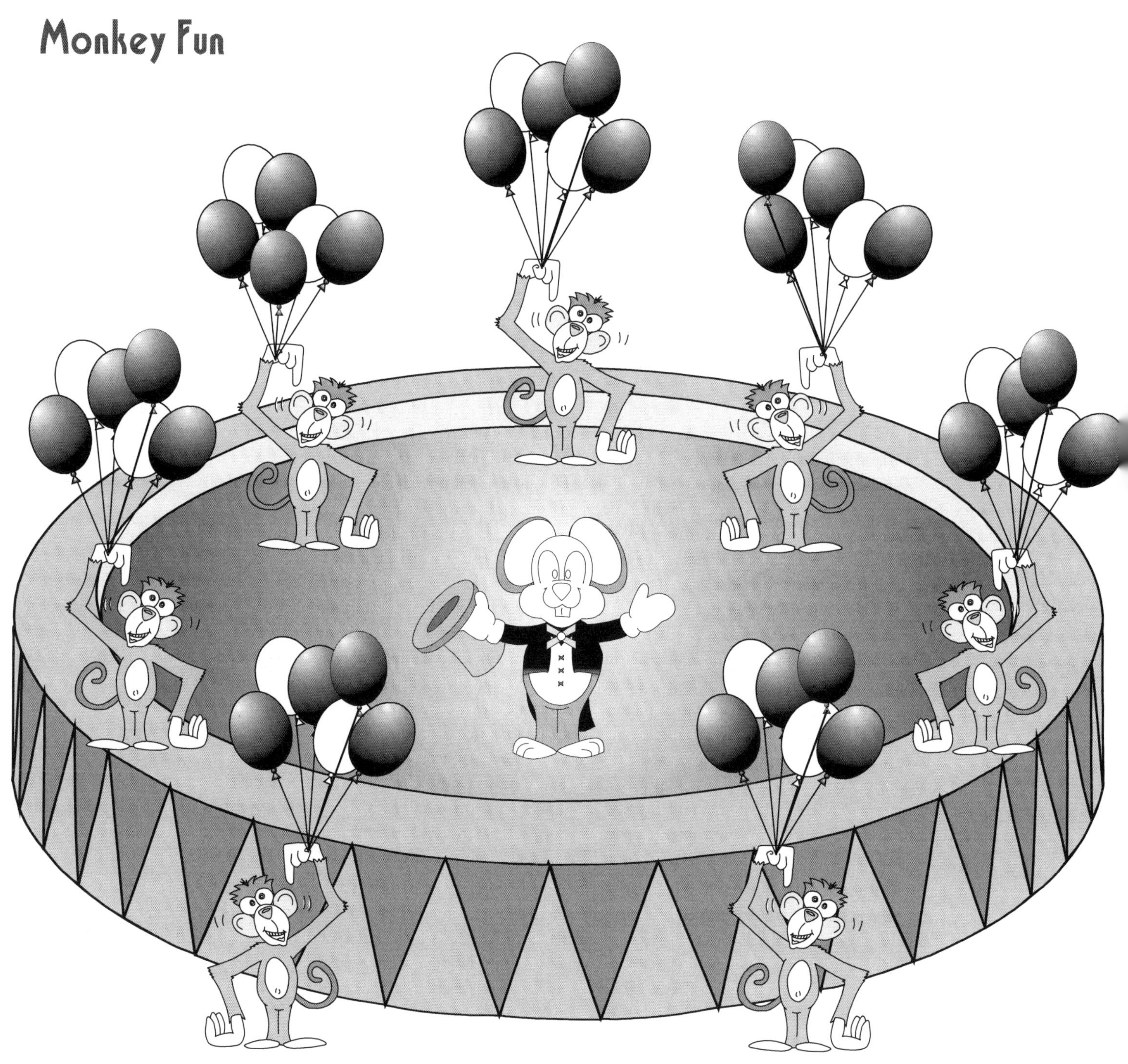

If all the balloons in this act pop, how many will Rudy need for the next show? Each monkey has the same number of balloons.

Let's help Rudy count the monkeys.
There are _____ monkeys.

Let's help Rudy count the balloons one monkey has.
Each monkey has _____ balloons.

_____ x _____=_____ balloons
Rudy will need _____ balloons.

Circus Snack Survey

Children voted for their favorite Circus Snacks. Here are the results:

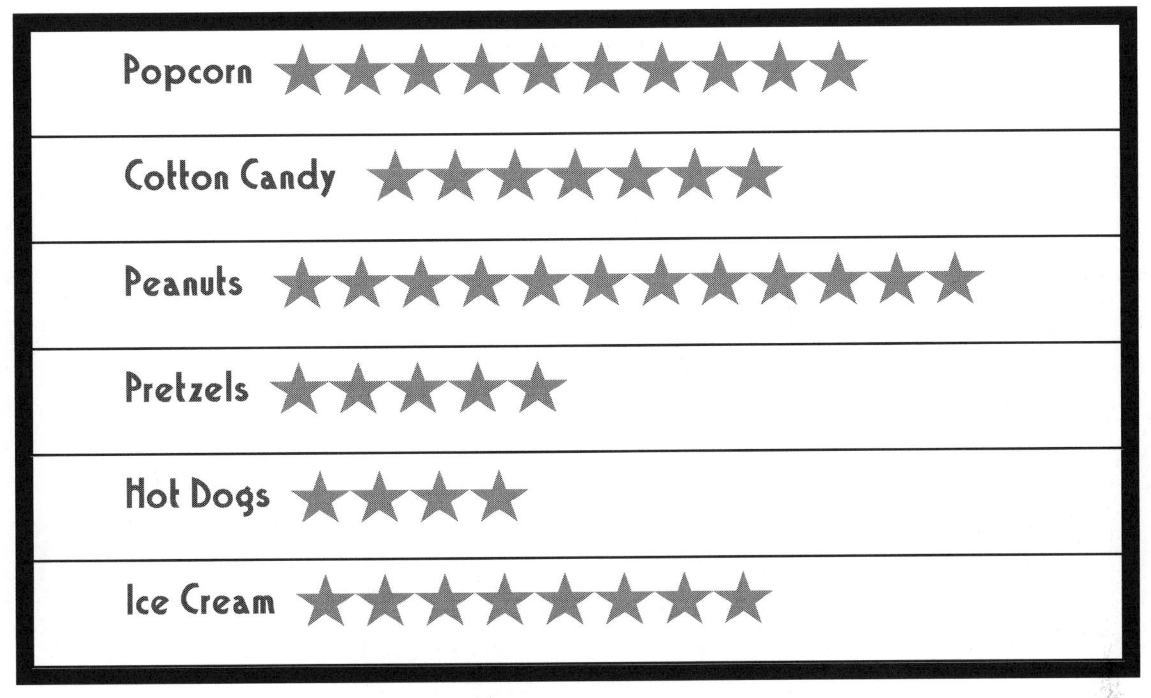

★ = 10 votes

1. Which snack got 50 votes? _________

2. How many votes did ice cream get? _________

3. Which snack got more votes than popcorn? _________

4. Which snack was the least favorite? _________

5. Which snack got 100 votes? _________

6. Which was the most popular snack? _________

7. Which of the above is your favorite snack? _________

The Trick to the 9's!

Part 1

Number **9** to **0** on the right.

X	9
1	0_
2	1_
3	2_
4	3_
5	4_
6	5_
7	6_
8	7_
9	8_
10	9_

Circus Tickets

Adults $5

Children $2

Anna bought 6 children's tickets. How much did she spend?

$2 x 6 = $12

Mrs. Robles bought 5 adult tickets. How much did she spend?

____ x ____ = $____

Coach bought 9 children's tickets and 6 adult tickets. How much did he spend?

$2 x 9 = $18

$5 x 6 = $30

$18
+$30
Total: $48

Joe's aunt bought the family 3 adult tickets and 7 children's tickets. How much did she spend?

____ x ____ = $____

____ x ____ = $____

Add the numbers.

Total: $____

Maria had a party at the circus. Her mother bought 4 adult tickets and 5 children's tickets. How much did they spend?

____ x ____ = $____

____ x ____ = $____

Add the numbers.

Total: $____

Betty spent $21 on circus tickets. She spent $15 on adult tickets How many adults were in her group? How many children were in the group?

$5 x ____ = $15

Adults: ____

$21
-$15
$6

$2 x ____ = $6

Children: ____

Trick to the 9's

Part 2

Number **0** to **9** on the left.

X	9
1	09
2	_8
3	_7
4	_6
5	_5
6	_4
7	_3
8	_2
9	_1
10	90

Decoding Patterns?

Write the multiplication problem for:

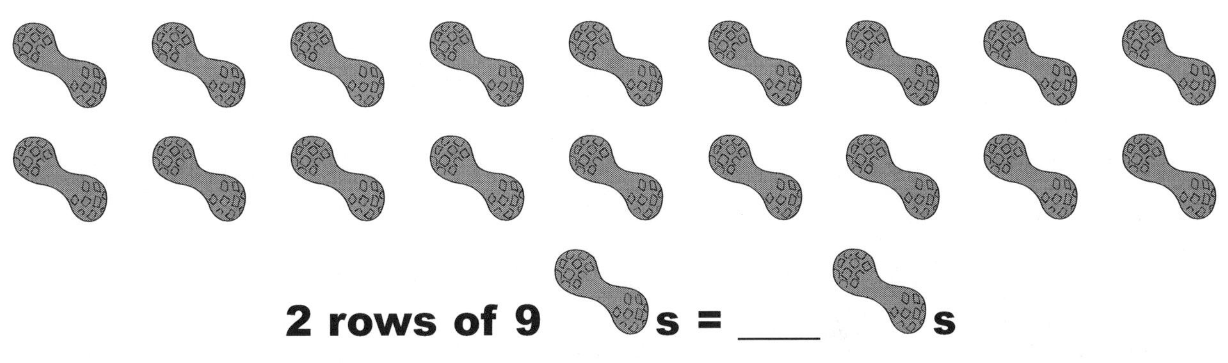

2 rows of 9 s = ____ s

2 x 9 = ____

even x odd = ____

___ rows of ___ s = ____ s

____ x ____ = ____

odd x ____ = ____

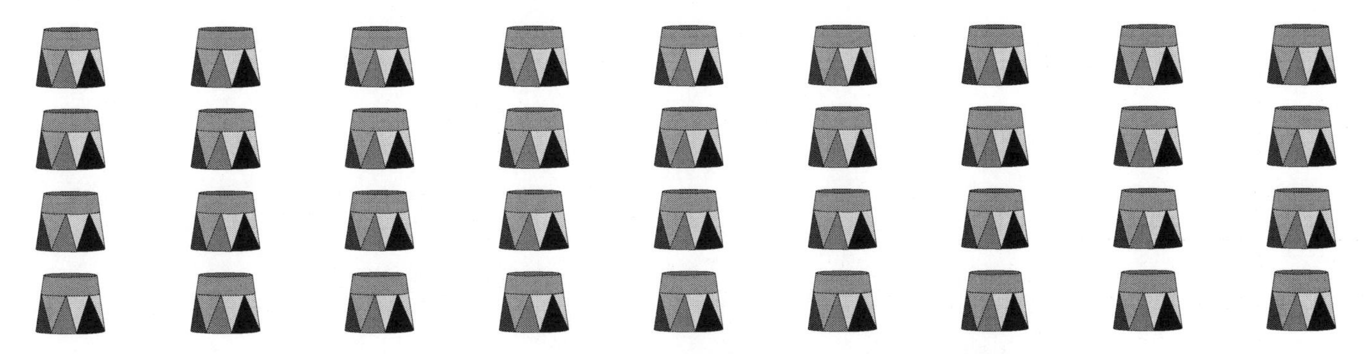

___ rows of ___ s = ____ s

___ x ___ = ____

even x ___ = ____

Magic 9's!

Separate the multiples into individual numbers and ADD. What did you discover?

X	9	Add	Total
1	9	0+9=	9
2	18	1+8=	
3	27		
4	36		
5	45		
6	54		
7	63		
8	72		
9	81		
10	90		

Magic Circus Fun

Help Rudy multiply the following.
Underneath, write the Odd/Even rule.

9 x 4 = ____
___ x ___ = even

3 x 9 = ____
___ x ___ = ____

8 x ___ = 72
___ x ___ = ____

2 x ___ = 18
___ x ___ = ____

7 x ___ = 63
___ x ___ = ____

5 x ___ = 45
___ x ___ = ____

9 x 7 = ____
___ x ___ = ____

9 x 10 = ____
___ x ___ = ____

9 x ___ = 54
___ x ___ = ____

9 x ___ = 27
___ x ___ = ____

9 x ___ = 81
___ x ___ = ____

9 x ___ = 36
___ x ___ = ____

Multiply:

9	9	9	9	8	4	2	6	10
x7	x3	x5	x9	x9	x9	x9	x9	x9

The 9's Flip Flop!

Flip flop the multiples.
What did you discover?

X	9		
1	09		90
2	18		81
3	27		
4	36		
5	45		
6	54		
7	63		
8	72		
9	81		
10	90		

Can you copy Leo on the blank grid?

Magic 9's!

Fill in the columns.

X	9	Add	Total
1	9	0+9=	9
2			
3			
4			
5			
6			
7			
8			
9			
10			

Help Bippo complete the pattern
and find his way home to
270 Magic Circus Drive.

Color the Clown

Solve the problems to color the clown.

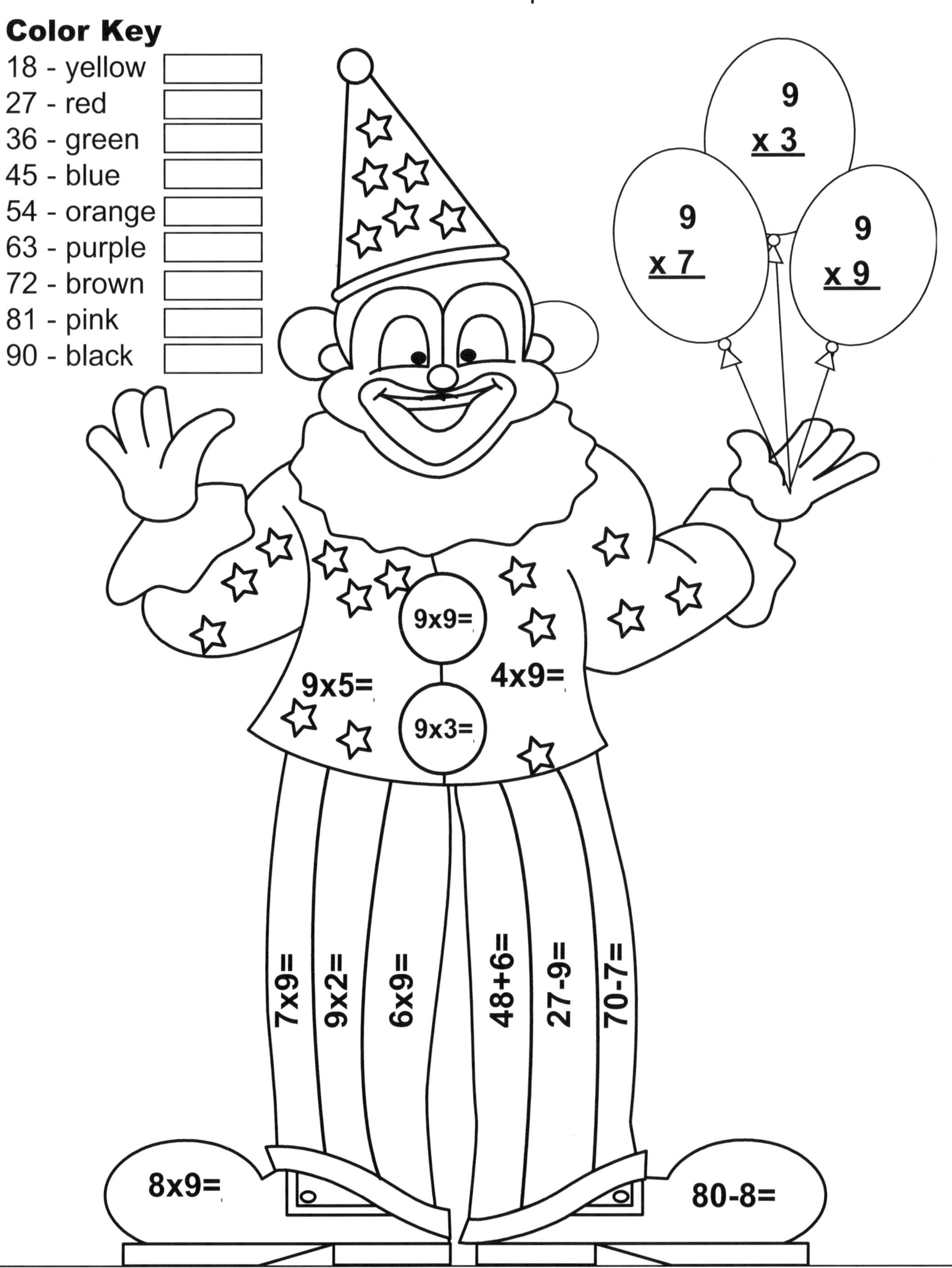

Mastering the 3's!

The 3 times table is super easy!
Fill in the _____ by adding 3
to the previous number.

X	3	Check it out!
1	3	0+3= 3
2	6	3+3= 6
3	9	6+3= 9
4	12	9+3= 12
5	15	12+3= 15
6	18	15+3= 18
7	21	18+3= 21
8	24	21+3= 24
9	27	24+3= 27
10	30	27+3= 30

Trick to the 3's!
3-6-9, we're doing fine!

Can you discover the trick to the 3's?

X	3	Add	Total
1	3	3	3
2	6	6	6
3	9	9	9
4	12	1+2=	
5	15		
6	18		
7	21		
8	24		
9	27		
10	30		

Magic Circus Fun

Help Rudy solve the following:

There are ____ giraffes in each ring.
How many giraffes in total?

____ x ____ = ____

There are ____ monkeys in each ring.
How many monkeys in total?

____ x ____ = ____

There are ____ lions in each ring.
How many lions in total?

____ x ____ = ____

Table 3 is as easy as 1, 2, 3!

What happens when you group table 3 in 3's?
Notice the 1 - 9 pattern?
How fun is that!

Fill in the ____ in each column.

3	__2	__1
6	__5	__4
9	__8	__7

(3 x 10 = 30)

What did you discover?
You filled in the center column with _____ and the column on the right with _____.

Let's try once more!
Fill in the ____ in each column.

3	1__	2__
6	1__	2__
9	1__	2__

Help Leo complete the pattern
and find his way home to
102 Magic Circus Drive.

3 6 9

30

60

Leo

102

90

Cool Trick to the 6's!

Although 6 is an EVEN number,
it is a multiple of 3.
It too must have a trick. Can you discover it?
(For 6 x 8, there are 2 steps.)

X	6	Add	Total
1	6	0+6=	6
2	12	1+2=	
3	18	1+8=	
4	24	2+4=	
5	30	3+0=	
6	36	3+6=	
7	42	4+2=	
8	48	4+8=12 1+2=	
9	54	5+4=	
10	60	6+0=	

Circus Snack Survey

Children voted for their favorite Circus Snacks. Here are the results:

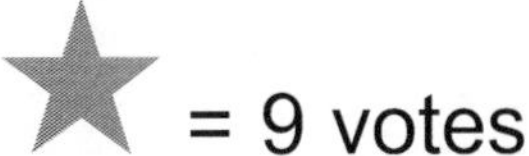

1. Which snack got 90 votes? _________

2. How many votes did hot dogs get? _________

3. Which snacks got more votes than peanuts? _________

4. Which snack got 63 votes? _________

5. Which was the most popular snack? _________

6. Which snack was the least favorite? _________

7. Which of the above is your favorite snack? _________

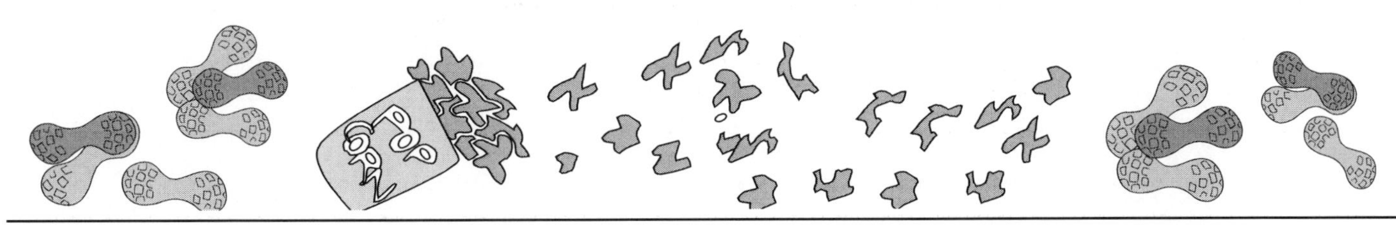

3, 6 and 9 Magic!

Fill in tables 3, 6 and 9.
Notice the hopscotch pattern of ODD multiples.

X	1	2	3	4	5	6	7	8	9	10
1										
2										
3	3	6	9	12	15	18	21	24	27	30
4										
5										
6	6	12	18	24	30	36	42	48	54	60
7										
8										
9	9	18	27	36	45	54	63	72	81	90
10										

Tricks of the 3, 6 and 9 Tables!
3-6-9 6-3-9 9-9-9

A super easy chart to help you remember!

X	3	Add	6	Add	9	Add
1	3	0+3=3	6	0+6=6	9	0+9=9
2	6	0+6=6	12	1+2=3	18	1+8=9
3	9	0+9=9	18	1+8=9	27	2+7=9
4	12	1+2=3	24	2+4=6	36	3+6=9
5	15	1+5=6	30	3+0=3	45	4+5=9
6	18	1+8=9	36	3+6=9	54	5+4=9
7	21	2+1=3	42	4+2=6	63	6+3=9
8	24	2+4=6	48	4+8=12 1+2=3	72	7+2=9
9	27	2+7=9	54	5+4=9	81	8+1=9
10	30	3+0=3	60	6+0=6	90	9+0=9

3, 6 and 9 Magic Review

Fill in tables 3, 6 and 9.

X	1	2	3	4	5	6	7	8	9	10
1										
2										
3										
4										
5										
6										
7										
8										
9										
10										

Circus Snacks

Item	Price	Item	Price
Hot Dogs	$3	Ice Cream	$2
Pretzels	$1	Popcorn	$3
Cotton Candy	$2	Soda	$1

Jim bought 6 bags of popcorn at the circus for his friends. How much did he spend?

_____ x _____ = $_____

Maria bought her sisters 5 pretzels. How much did she spend?

_____ x _____ = $_____

Sue's sister bought 2 cotton candies and 3 sodas. How much did she spend?

_____ x _____ = $_____

_____ x _____ = $_____

Total:
_____ + _____ = $_____

Coach bought the team 9 hot dogs and 10 sodas. How much did he spend?

_____ x _____ = $_____

_____ x _____ = $_____

Total:
_____ + _____ = $_____

Bill and Tom were hungry. They bought 6 hot dogs, 4 sodas and 3 ice creams. How much did they spend?

_____ x _____ = $_____

_____ x _____ = $_____

_____ x _____ = $_____

Total:
_____ +_____ + _____ = $_____

Alicia bought her family 5 sodas, 6 pretzels and 7 hot dogs. How much did she spend?

_____ x _____ = $_____

_____ x _____ = $_____

_____ x _____ = $_____

Total:
_____ +_____ + _____ = $_____

One More Time with 3, 6 and 9!

Look at the examples and fill in the chart.
Don't forget the 2 steps with 48. ***Good job!***

X	3	Add	6	Add	9	Add
1	3	0+3=	6	0+6=	9	0+9=
2	6	0+6=	12		18	
3	9	0+9=	18		27	
4	12		24		36	
5	15		30		45	
6	18		36		54	
7	21		42		63	
8	24		48		72	
9	27		54		81	
10	30		60		90	

Help Rudy multiply the following.
Underneath, write the Odd/Even rule.

___ x ___ = ___
___ x ___ = EVEN

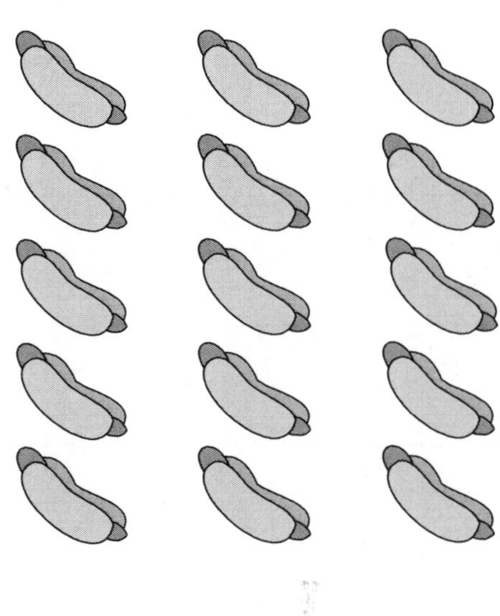

___ x ___ = ___
___ x ___ = ODD

___ x ___ = ___
___ x ___ = ___

___ x ___ = ___
___ x ___ = ___

Odd or Even Multiple?

Secret Clue: If one number is **even**, the multiple is **even**.
Circle every **even** number first. Fill in with **odd** or **even**.

6 x 3 = even	5 x 9 = odd	8 x 4 = even
3 x 7 = ____	4 x 6 = ____	2 x 3 = ____
5 x 3 = ____	9 x 2 = ____	4 x 5 = ____
1 x 9 = ____	6 x 4 = ____	8 x 7 = ____
9 x 6 = ____	3 x 5 = ____	2 x 8 = ____
6 x 5 = ____	2 x 7 = ____	8 x 5 = ____
4 x 7 = ____	3 x 9 = ____	4 x 3 = ____
5 x 2 = ____	4 x 1 = ____	5 x 6 = ____
8 x 1 = ____	3 x 4 = ____	9 x 3 = ____
6 x 7 = ____	6 x 8 = ____	8 x 2 = ____
3 x 8 = ____	8 x 9 = ____	5 x 5 = ____
6 x 2 = ____	4 x 3 = ____	9 x 1 = ____

Color the Clown

Solve the problems to color the clown.

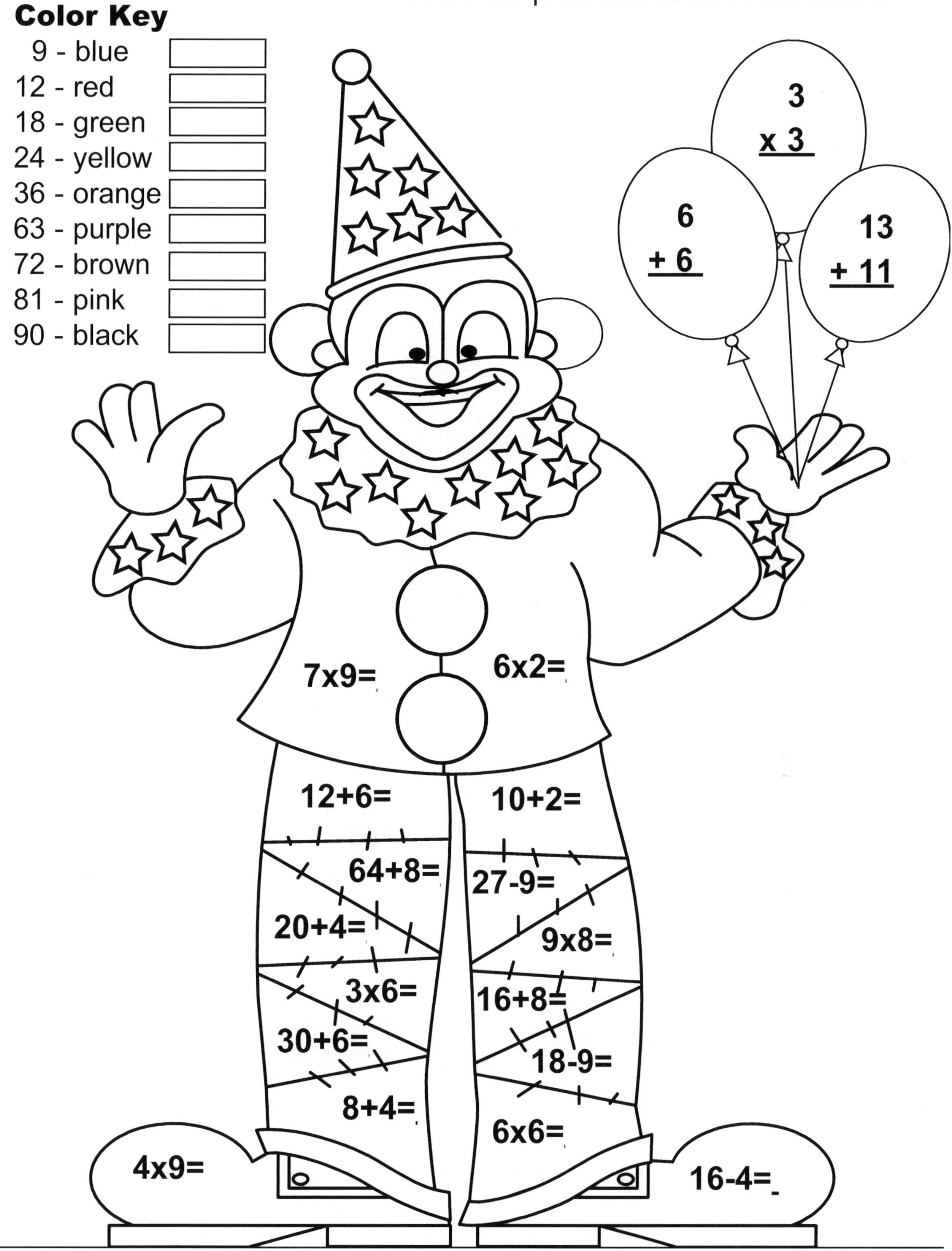

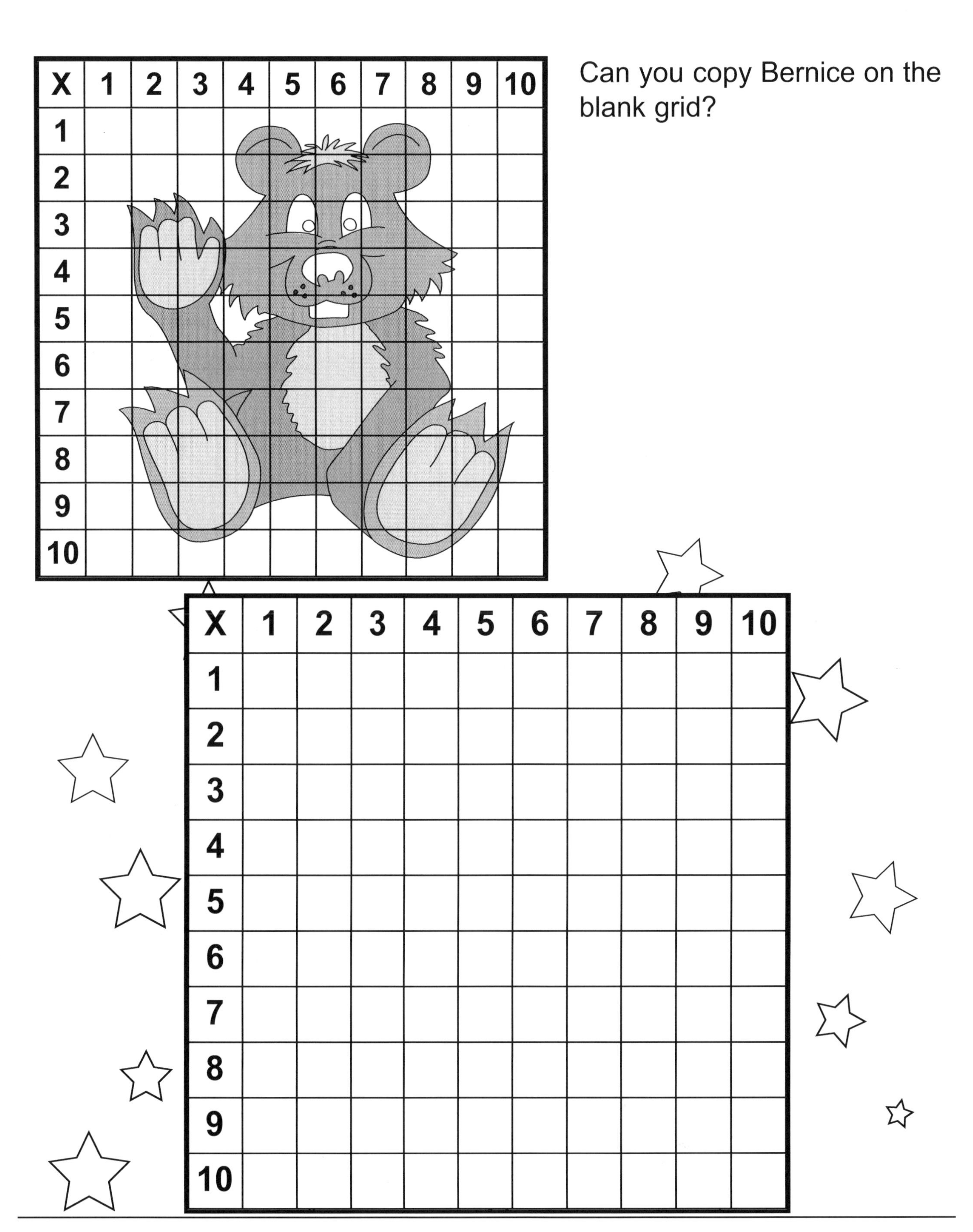

Can you copy Bernice on the blank grid?

Amazing Clue to the 7's!

The 3 x table holds the clue to table 7.
Take the underlined digit in the 3 x table ***reversed***
and fill in the 7 x table. ***Pretty amazing!.***

X		*3 x table Reversed*	7
0	*10*	*30*	0
1	*9*	*27*	_
2	*8*	*24*	1_
3	*7*	*21*	2_
4	*6*	*18*	2_
5	*5*	*15*	3_
6	*4*	*12*	4_
7	*3*	*9*	4_
8	*2*	*6*	5_
9	*1*	*3*	6_
10	*0*	*0*	7_

Amazing Clue Review

Fill in the blanks.

X		3 x table Reversed	7
0	10	30	0
1	9	27	7
2	8	2_	1_
3	7	2_	2_
4	6	1_	2_
5	5	1_	3_
6	4	1_	4_
7	3	_	4_
8	2	_	5_
9	1	_	6_
10	0	0	7_

Odd or Even?

Help Rudy group in pairs. Example: 

Any ODD [monkey], [ticket], [stand], [peanut] left over?
Write a multiplication problem for each.

__ x __ = __

__ x __ = __

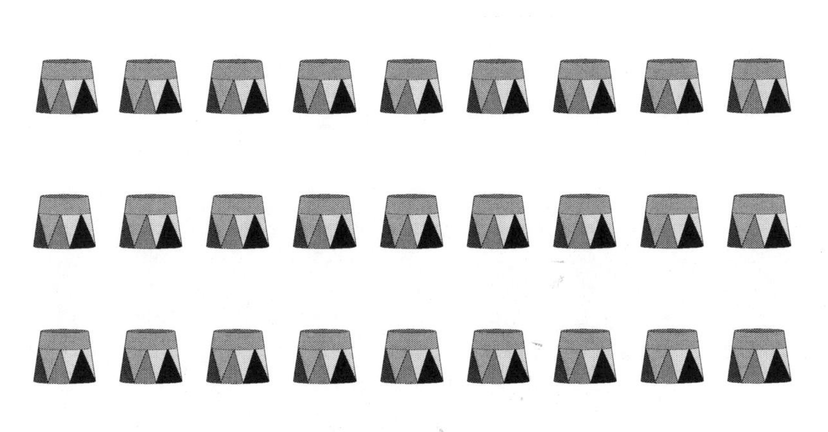

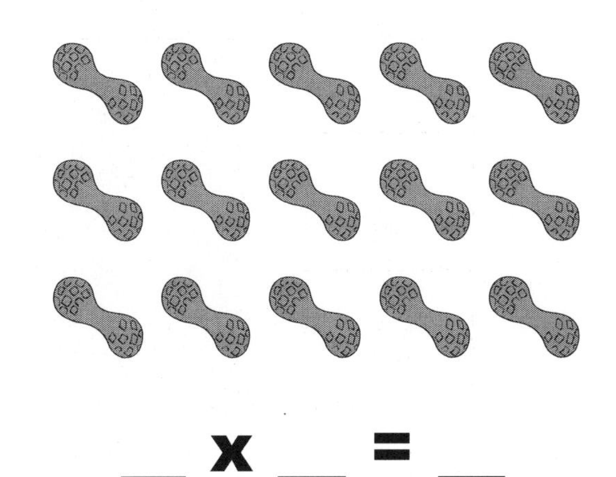

__ x __ = __

__ x __ = __

7's Challenge!

Fill in the 7 x table.

X		3 x table Reversed	7
0	10	30	0
1	9	27	__
2	8	24	__
3	7	21	__
4	6	18	__
5	5	15	__
6	4	12	__
7	3	9	__
8	2	6	__
9	1	3	__
10	0	0	70

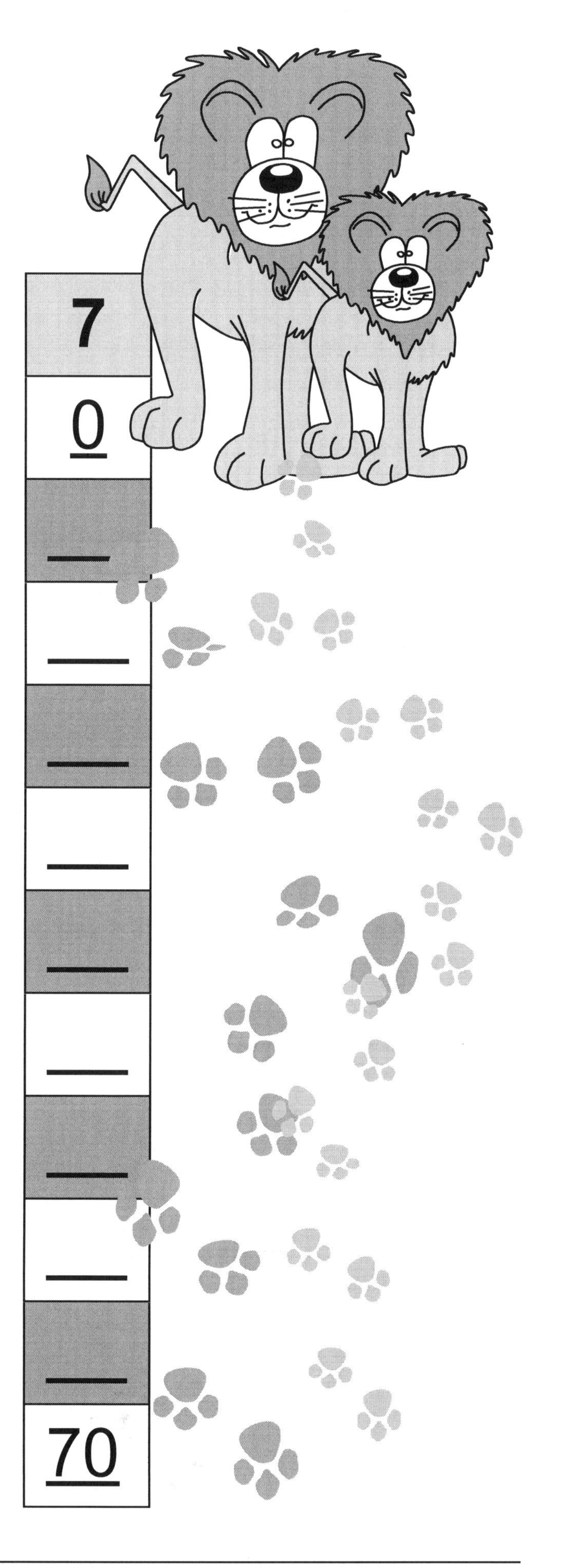

Clown Fun

Fill in the blanks.

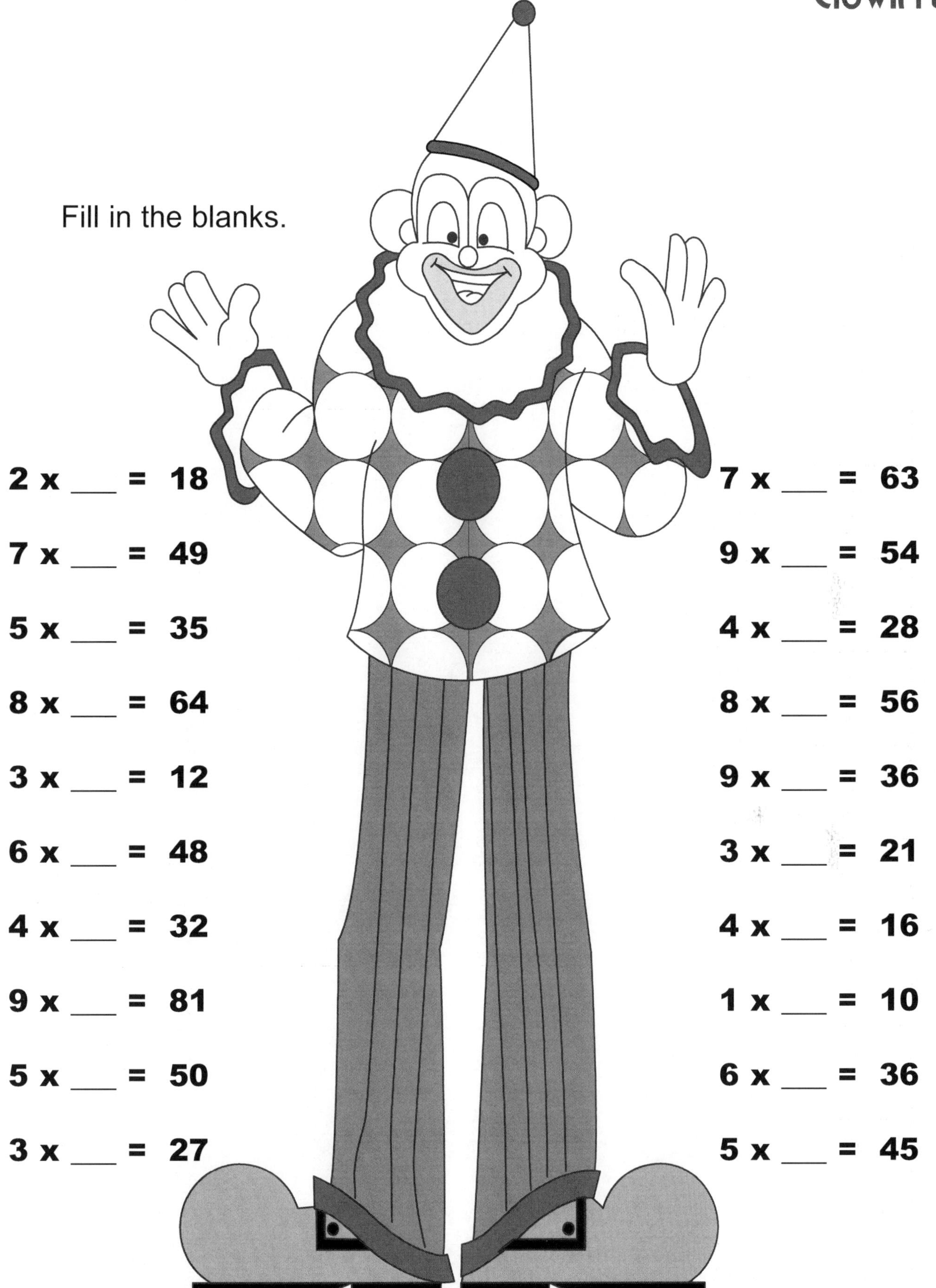

2 x ___ = 18

7 x ___ = 49

5 x ___ = 35

8 x ___ = 64

3 x ___ = 12

6 x ___ = 48

4 x ___ = 32

9 x ___ = 81

5 x ___ = 50

3 x ___ = 27

7 x ___ = 63

9 x ___ = 54

4 x ___ = 28

8 x ___ = 56

9 x ___ = 36

3 x ___ = 21

4 x ___ = 16

1 x ___ = 10

6 x ___ = 36

5 x ___ = 45

Table 7 is as easy as 1, 2, 3!

What happens when you group table 7 in 3's?
Notice the 1-9 pattern?
How fun is that!

Fill in the ____ in each column.

7	__8	__9
__4	__5	__6
__1	__2	__3

(7 x 10 = 70)

Let's try once more!
Fill in the ____ in each column.

7	2__	4__
1__	3__	5__
2__	4__	6__

Missing Evens!

Fill in the blanks.

3 x ___ = 24	**8 x ___ = 64**	**6 x ___ = 24**
6 x ___ = 12	**5 x ___ = 30**	**7 x ___ = 28**
1 x ___ = 10	**9 x ___ = 72**	**5 x ___ = 20**
3 x ___ = 18	**4 x ___ = 16**	**3 x ___ = 30**
2 x ___ = 16	**5 x ___ = 40**	**5 x ___ = 10**
7 x ___ = 14	**9 x ___ = 36**	**9 x ___ = 54**
5 x ___ = 50	**4 x ___ = 32**	**2 x ___ = 20**
8 x ___ = 48	**9 x ___ = 18**	**6 x ___ = 48**
4 x ___ = 24	**8 x ___ = 32**	**7 x ___ = 42**
8 x ___ = 16	**7 x ___ = 56**	**4 x ___ = 40**
3 x ___ = 12	**8 x ___ = 80**	**6 x ___ = 36**
9 x ___ = 90	**Good Job!**	**7 x ___ = 70**

Missing Odds!

Fill in the blanks.

7 x ___ = 21	3 x ___ = 15	9 x ___ = 81
5 x ___ = 35	6 x ___ = 30	2 x ___ = 18
6 x ___ = 54	8 x ___ = 72	5 x ___ = 45
4 x ___ = 28	2 x ___ = 10	8 x ___ = 40
3 x ___ = 21	4 x ___ = 20	6 x ___ = 42
2 x ___ = 14	7 x ___ = 49	9 x ___ = 45
8 x ___ = 24	9 x ___ = 27	7 x ___ = 63
6 x ___ = 18	4 x ___ = 36	5 x ___ = 15
3 x ___ = 27	8 x ___ = 56	7 x ___ = 35
4 x ___ = 12	6 x ___ = 6	9 x ___ = 63
5 x ___ = 25	**Good Job!**	10 x ___ = 90

Help Barry complete the pattern and find his way home to **175 Magic Circus Drive.**

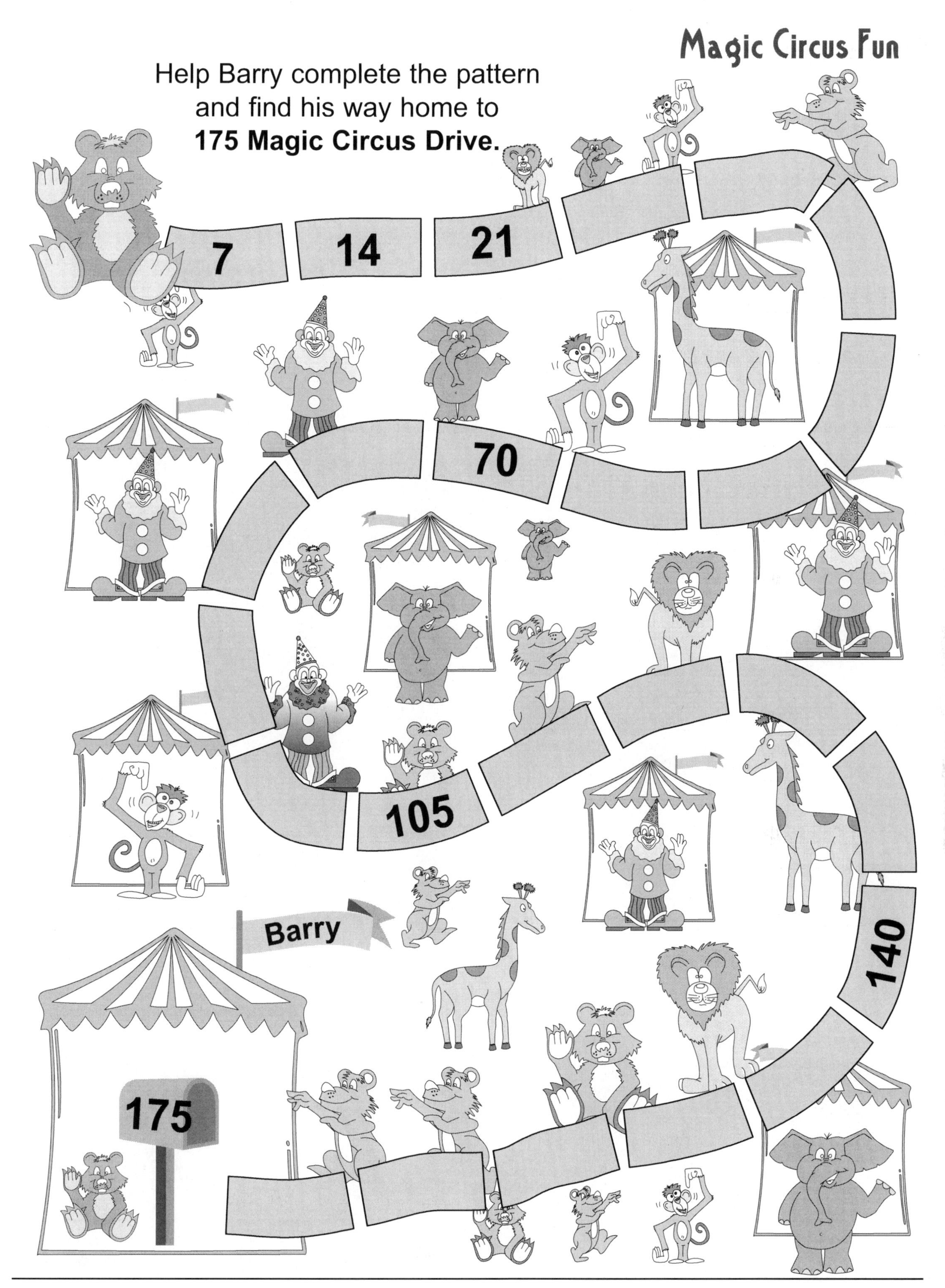

Fun Patterns

All ODD number multiples end in 1-9
when you multiply 1 through 9
except table 5 which has a 5-0-5-0 pattern.

Fill in table 9:

9	___6	___3
___8	___5	___2
___7	___4	___1

Fill in table 1:

1	4	7
___	___	___
___	___	___

Fill in table 7:

7	___8	___9
___4	___5	___6
___1	___2	___3

Fill in table 3:

3	___2	___1
6	___5	___4
9	___8	___7

What Are the Odds?

Fill in tables 1-10.
Notice the **hopscotch** pattern for **ODD** numbers.
Out of 100 multiples only _____ were ODD.
How odd is that?

Odd x Odd = ODD
Even x ANY number = EVEN

X	1	2	3	4	5	6	7	8	9	10
1										
2										
3										
4										
5										
6										
7										
8										
9										
10										

Monkey Fun

If all the balloons in this act pop, how many will Rudy need for the next show? Each monkey has the same number of balloons.

Let's help Rudy count the monkeys.
There are _____ monkeys.

Let's help Rudy count the balloons one monkey has.
Each monkey has ____ balloons.

____ x _____ = ____ balloons
Rudy will need ____ balloons.

Squaring a number is multiplying the number times itself.
Notice the square formed by 2 x 2, 3 x 3, 4 x 4 and 5 x 5.
Pretty cool!

X	1	2	3	4	5
1	•	••	•••	••••	•••••
2	• •	•• ••	••• •••	•••• ••••	••••• •••••
3	• • •	•• •• ••	••• ••• •••	•••• •••• ••••	••••• ••••• •••••
4	• • • •	•• •• •• ••	••• ••• ••• •••	•••• •••• •••• ••••	••••• ••••• ••••• •••••
5	• • • • •	•• •• •• •• ••	••• ••• ••• ••• •••	•••• •••• •••• •••• ••••	••••• ••••• ••••• ••••• •••••

Fun Squaring Numbers

A number multiplied times itself creates a square on the grid.
Start with the single SQUARE of 1 x 1 and step by step
go down the 100 square grid of 10 x 10.
Notice how the squares get larger and larger.
Outline each additional square a different color.
How many squares did you end up with? ____

X	1	2	3	4	5	6	7	8	9	10
1	1									
2		4								
3			9							
4				16						
5					25					
6						36				
7							49			
8								64		
9									81	
10										100

Here's the Answer!

What's the Problem?

Fill in the blanks. If an answer appears twice, give another possibility using different numbers. Do not multiply by 1.

Example: **3 x 4** = 12 or **6 x 2** = 12

___ x ___ = 40	___ x ___ = 40	___ x ___ = 18
___ x ___ = 18	___ x ___ = 9	___ x ___ = 15
___ x ___ = 24	___ x ___ = 24	___ x ___ = 16
___ x ___ = 16	___ x ___ = 10	___ x ___ = 48
___ x ___ = 30	___ x ___ = 25	___ x ___ = 36
___ x ___ = 36	___ x ___ = 64	___ x ___ = 14
___ x ___ = 63	___ x ___ = 72	___ x ___ = 35
___ x ___ = 81	___ x ___ = 27	___ x ___ = 42
___ x ___ = 32	___ x ___ = 60	___ x ___ = 6
___ x ___ = 54	___ x ___ = 56	___ x ___ = 21
___ x ___ = 45	___ x ___ = 28	___ x ___ = 50
___ x ___ = 12	___ x ___ = 12	___ x ___ = 0
___ x ___ = 20	**Good Job!**	___ x ___ = 20

You Know the Odds!

Fill in tables 1, 3, 5, 7 & 9.
Notice odd/even **hopscotch** pattern.

X	1		3		5		7		9
1	___		___		___		___		___
2	___		___		1_		1_		1_
3	___		___		1_		2_		2_
4	___		1_		2_		2_		3_
5	___		1_		2_		3_		4_
6	___		1_		3_		4_		5_
7	___		2_		3_		4_		6_
8	___		2_		4_		5_		7_
9	___		2_		4_		6_		8_
10	1_		3_		5_		7_		9_

Solve the Missing Diagonal

Fill in the missing diagonal.
Remember **Odd x Odd = Odd.**
Even x Even = Even.

X	1	2	3	4	5	6	7	8	9	10
1		2	3	4	5	6	7	8	9	10
2	2		6	8	10	12	14	16	18	20
3	3	6		12	15	18	21	24	27	30
4	4	8	12		20	24	28	32	36	40
5	5	10	15	20		30	35	40	45	50
6	6	12	18	24	30		42	48	54	60
7	7	14	21	28	35	42		56	63	70
8	8	16	24	32	40	48	56		72	80
9	9	18	27	36	45	54	63	72		90
10	10	20	30	40	50	60	70	80	90	

Odd Number Diagonals

Fill in the diagonals.
Notice how the numbers repeat on either side.
Wow, it's like magic!

X	1	2	3	4	5	6	7	8	9	10
1	1		3		5					
2		4		8						
3	3		9							
4		8		16						
5	5				25					
6						36				
7							49			
8								64		
9									81	
10										100

Even Number Diagonals

Fill in the diagonals. What did you discover about diagonal patterns for EVEN numbers?

X	1	2	3	4	5	6	7	8	9	10
1		2		4						
2	2		6							
3		6								
4	4									
5										
6										
7										
8										
9										
10										

Odd or Even Multiple?

Secret Clue: If one number is **even**, the multiple is **even**.
Circle every **even** number first. Fill in with **odd** or **even**.

[4] x 3 = even	3 x 5 = odd	[6] x [2] = even
5 x 7 = ____	4 x 8 = ____	3 x 3 = ____
6 x 3 = ____	5 x 8 = ____	9 x 7 = ____
9 x 9 = ____	6 x 9 = ____	7 x 8 = ____
6 x 6 = ____	3 x 9 = ____	2 x 4 = ____
5 x 5 = ____	2 x 6 = ____	8 x 1 = ____
4 x 7 = ____	7 x 5 = ____	2 x 3 = ____
9 x 2 = ____	4 x 4 = ____	5 x 9 = ____
3 x 1 = ____	3 x 8 = ____	7 x 3 = ____
7 x 7 = ____	6 x 8 = ____	8 x 3 = ____
8 x 8 = ____	8 x 7 = ____	9 x 5 = ____
5 x 2 = ____	4 x 5 = ____	7 x 1 = ____

Cool Diagonals!

With a ruler, draw diagonals using a different color for each.
Notice for **EVEN** numbers, the pattern ***duplicates*** itself.
The diagonal for **4** (which has **4** numbers) is: **4, 6, 6, 4.**

For **ODD** numbers, a number in the **middle** breaks the pattern.
The diagonal for **5** (which has **5** numbers) is: 5, 8, 9, 8, 5.

How many numbers in the diagonals for 3, 6, 7, 8, 9 and 10?
Pretty cool!

X	1	2	3	4	5	6	7	8	9	10
1	1	2	3	4	5	6	7	8	9	10
2	2	4	6	8	10	12	14	16	18	20
3	3	6	9	12	15	18	21	24	27	30
4	4	8	12	16	20	24	28	32	36	40
5	5	10	15	20	25	30	35	40	45	50
6	6	12	18	24	30	36	42	48	54	60
7	7	14	21	28	35	42	49	56	63	70
8	8	16	24	32	40	48	56	64	72	80
9	9	18	27	36	45	54	63	72	81	90
10	10	20	30	40	50	60	70	80	90	100

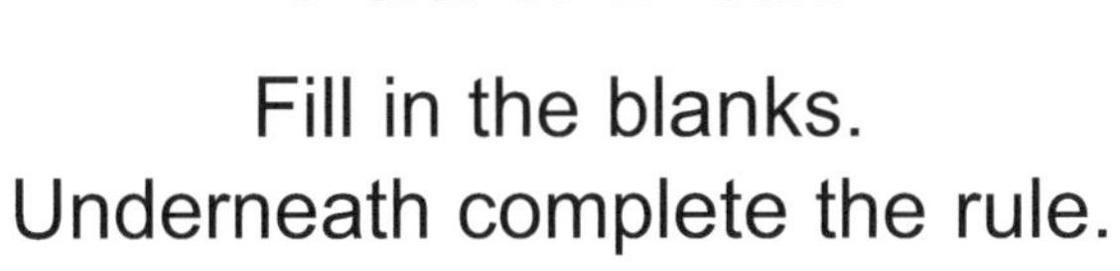

Odd or Even?

Fill in the blanks.
Underneath complete the rule.

8 x <u>4</u> = 32	**7 x ___ = 21**	**9 x ___ = 54**
e x <u>e</u> = e	**o x ___ = o**	**o x ___ = e**
3 x __ = 24	**5 x ___ = 25**	**2 x ___ = 18**
o x __ = e	**o x ___ = o**	**e x ___ = e**
9 x __ = 63	**7 x ___ = 28**	**4 x ___ = 20**
o x __ = o	**o x ___ = e**	**e x ___ = e**
6 x __ = 42	**4 x ___ = 36**	**5 x ___ = 45**
e x __ = e	**e x ___ = e**	**o x ___ = o**
3 x __ = 27	**8 x ___ = 64**	**9 x ___ = 81**
o x __ = o	**e x ___ = e**	**o x ___ = o**

Diagonal Flip Flop

Fill in the diagonals.
What did you discover?
Multiplication is super easy!

X	1	2	3	4	5	6	7	8	9	10
1	1	2	3	4						
2	2	4	6							
3	3	6	9							
4	4	8	12	16						
5	5	10	15	20	25					
6	6	12	18	24	30	36				
7	7	14	21	28	35	42	49			
8	8	16	24	32	40	48	56	64		
9	9	18	27	36	45	54	63	72	81	
10	10	20	30	40	50	60	70	80	90	100

Grid Magic

Can you copy Coco on the blank grid?

Diagonal Flip Flop

Fill in the diagonals.
What did you discover?
Numbers are magic!

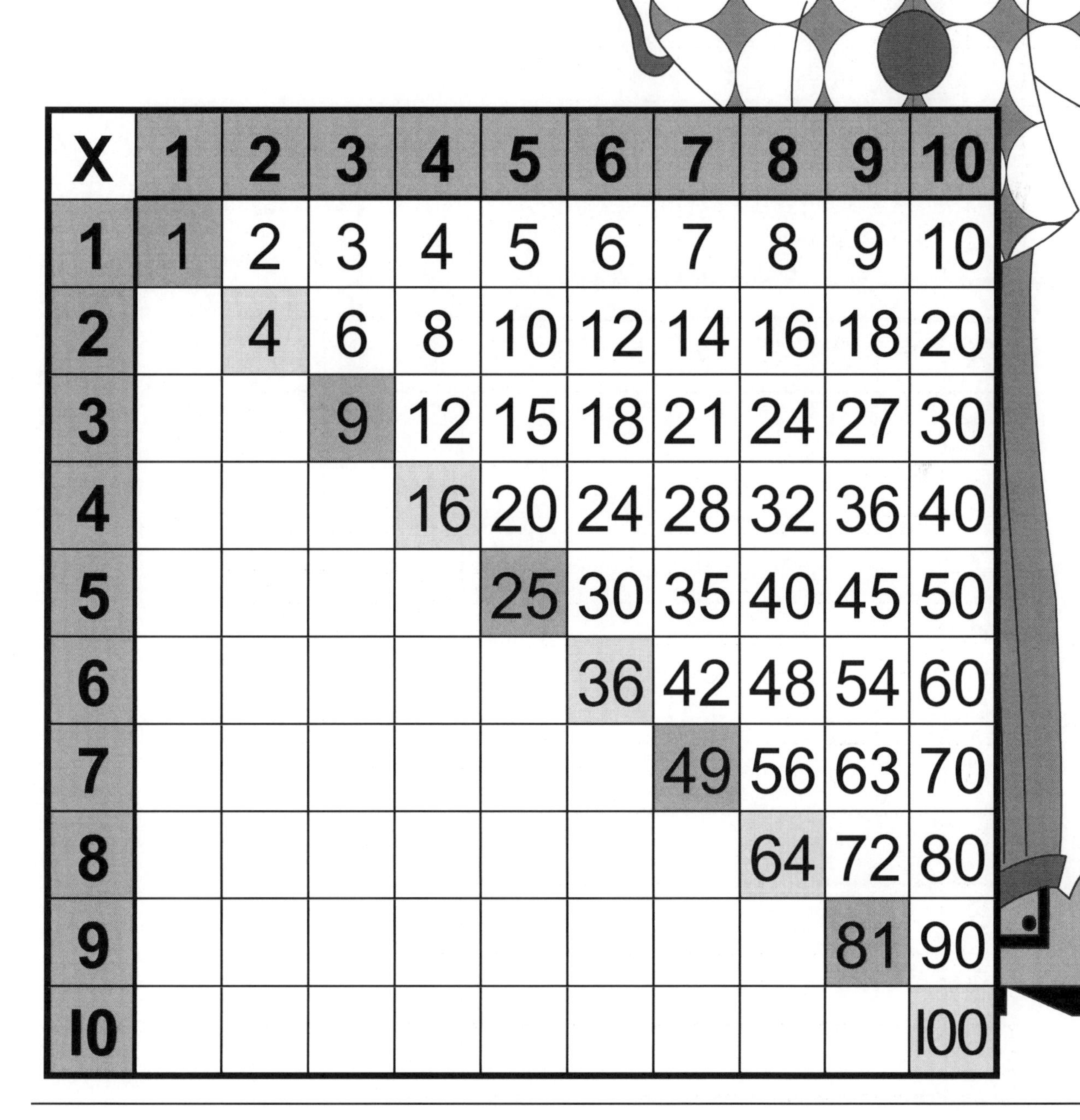

X	1	2	3	4	5	6	7	8	9	10
1	1	2	3	4	5	6	7	8	9	10
2		4	6	8	10	12	14	16	18	20
3			9	12	15	18	21	24	27	30
4				16	20	24	28	32	36	40
5					25	30	35	40	45	50
6						36	42	48	54	60
7							49	56	63	70
8								64	72	80
9									81	90
10										100

Grid Magic

Can you copy Rex on the blank grid?

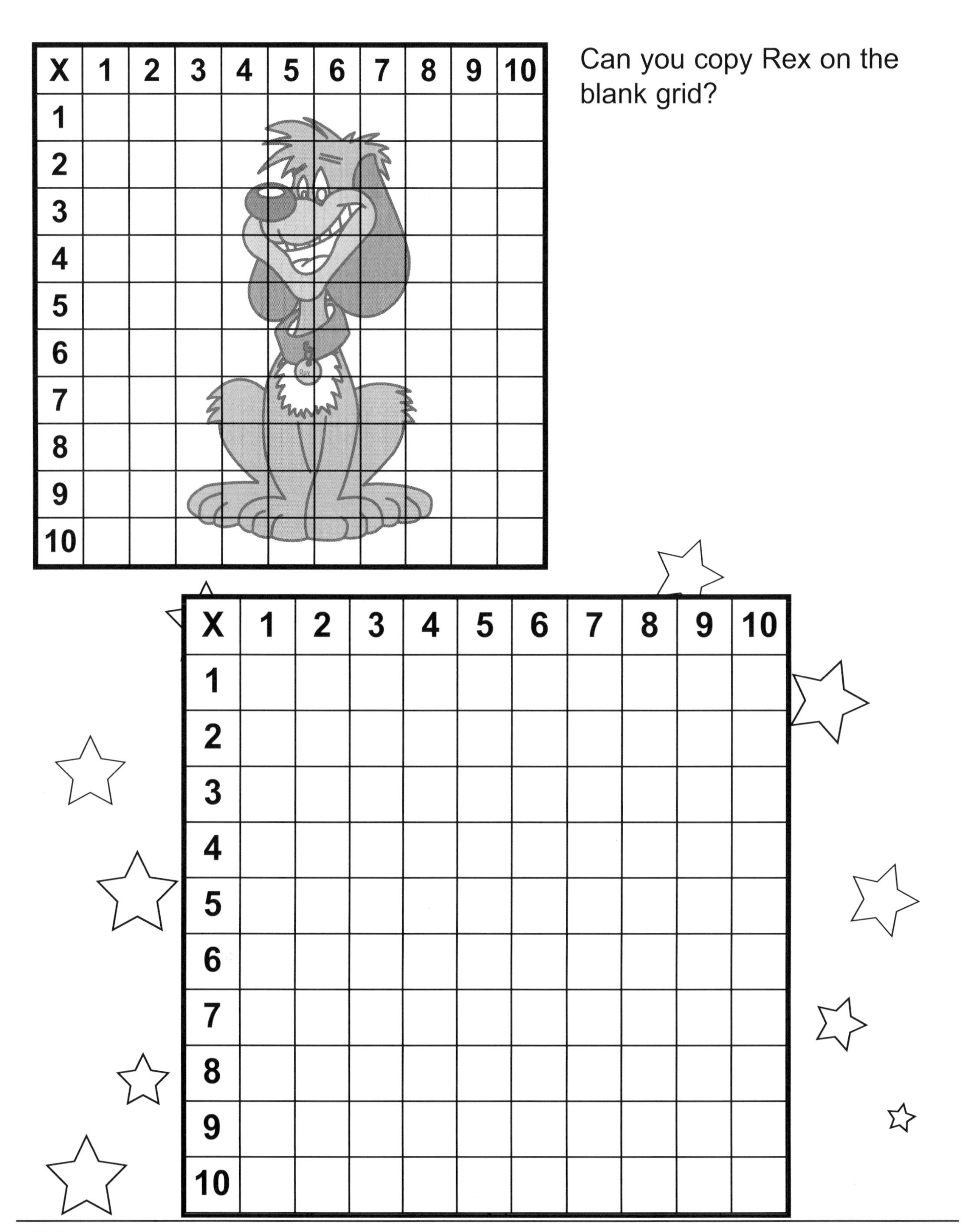

Super X Diagonals?

Fill in the diagonals.
Notice how the multiples in the top and left sides match.
So do the bottom and right sides.
Why is that? ***Pretty cool!***

X	1	2	3	4	5	6	7	8	9	10
1		2	3	4	5	6	7	8	9	
2	2		6	8	10	12	14	16		20
3	3	6		12	15	18	21		27	30
4	4	8	12		20	24		32	36	40
5	5	10	15	20			35	40	45	50
6	6	12	18	24			42	48	54	60
7	7	14	21		35	42		56	63	70
8	8	16		32	40	48	56		72	80
9	9		27	36	45	54	63	72		90
10		20	30	40	50	60	70	80	90	

Grid Magic

Can you copy Bippo on the blank grid?

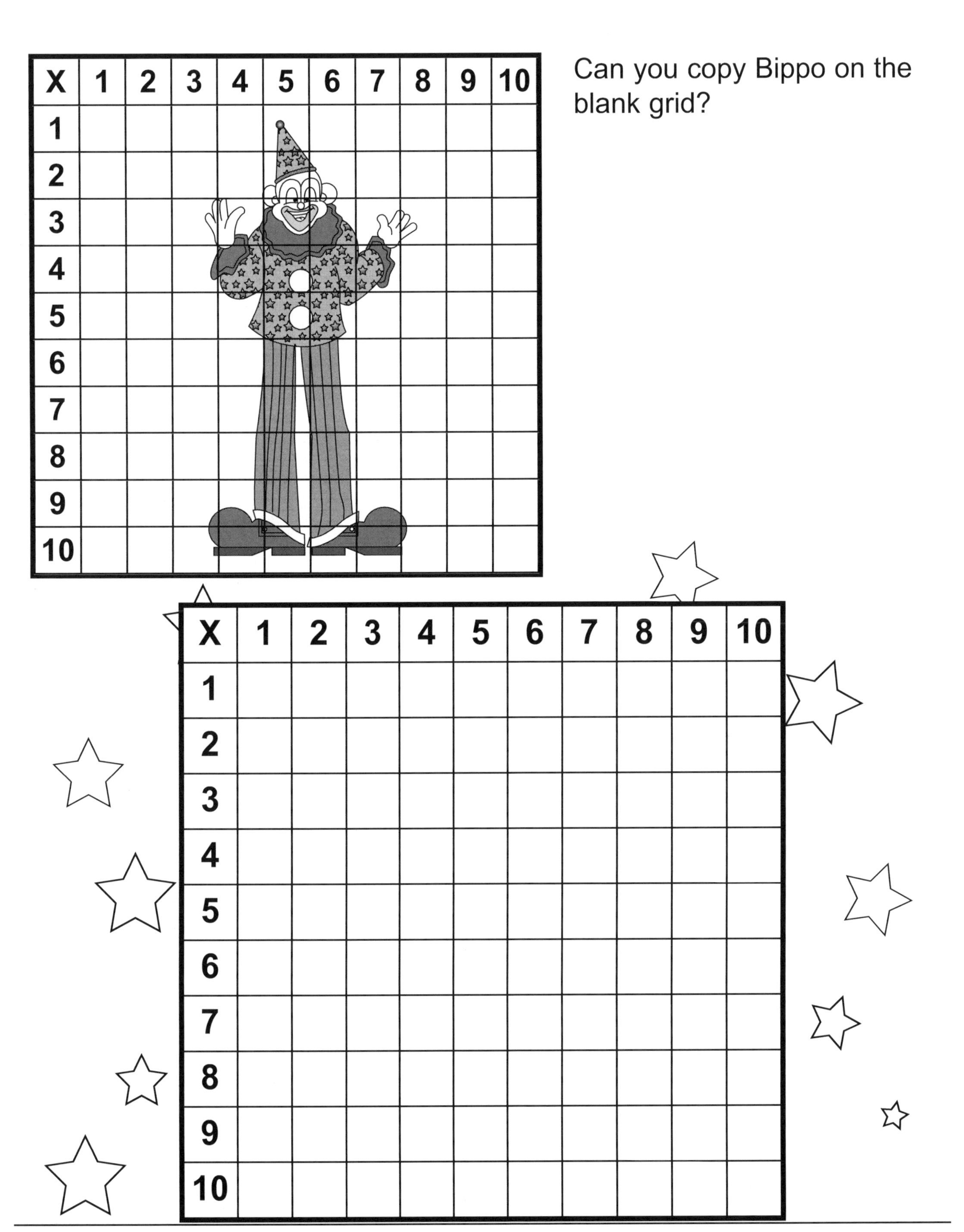

ODDS Diagonal Challenge!

Fill in the white squares. **Secret clue** - to work really fast, fill in the diagonal 1 - 100 **first** and then the multiples on **either side.**
You'll soon discover a super easy ***mirror pattern!***

X	1	2	3	4	5	6	7	8	9	10
1										
2										
3										
4										
5										
6										
7										
8										
9										
10										

Clown Fun

What's missing?

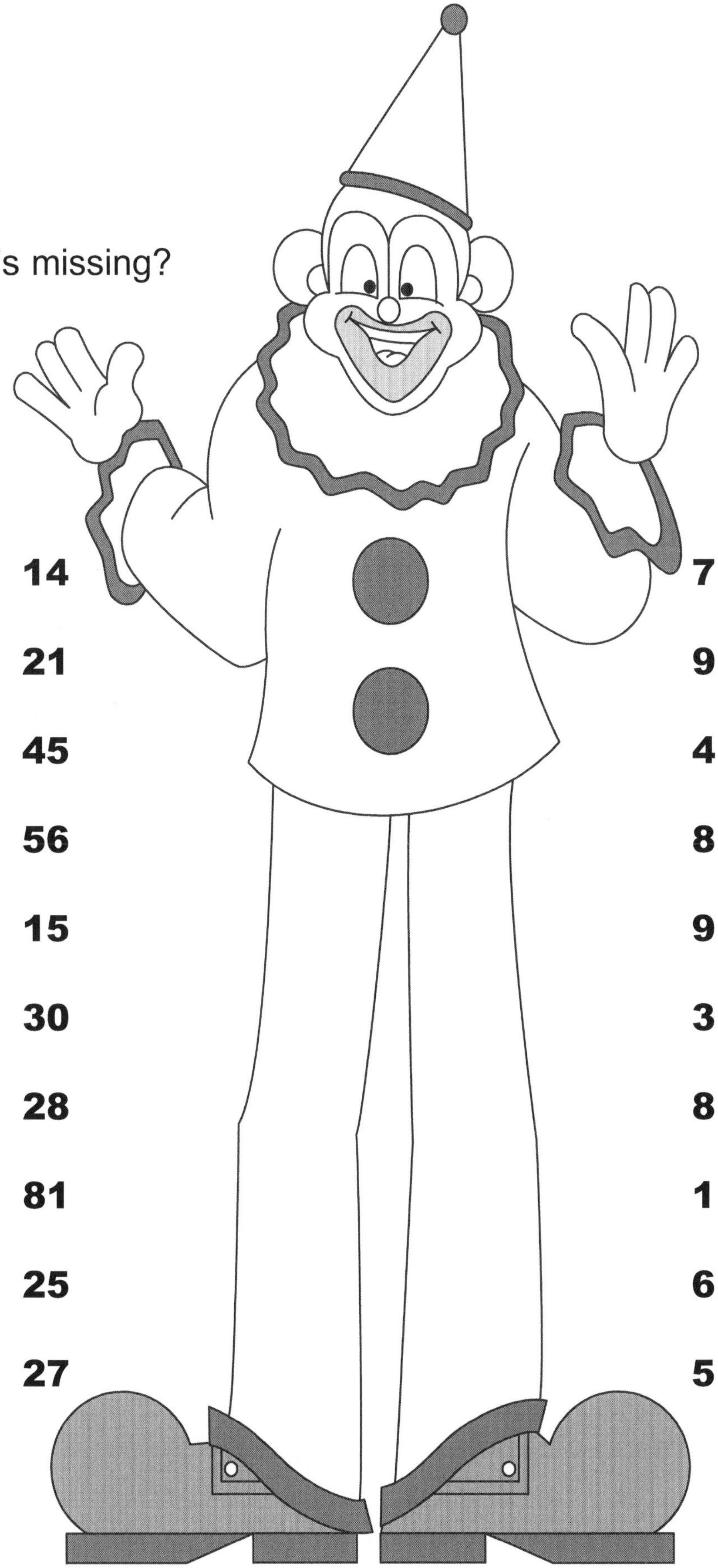

Missing ODDS

2 x ___ = 14

7 x ___ = 21

5 x ___ = 45

8 x ___ = 56

3 x ___ = 15

6 x ___ = 30

4 x ___ = 28

9 x ___ = 81

5 x ___ = 25

3 x ___ = 27

Missing EVENS

7 x ___ = 70

9 x ___ = 54

4 x ___ = 32

8 x ___ = 64

9 x ___ = 36

3 x ___ = 18

8 x ___ = 32

1 x ___ = 10

6 x ___ = 36

5 x ___ = 40

EVENS Diagonal Challenge!

Fill in the white squares.

X	1	2	3	4	5	6	7	8	9	10
1										
2										
3										
4										
5										
6										
7										
8										
9										
10										

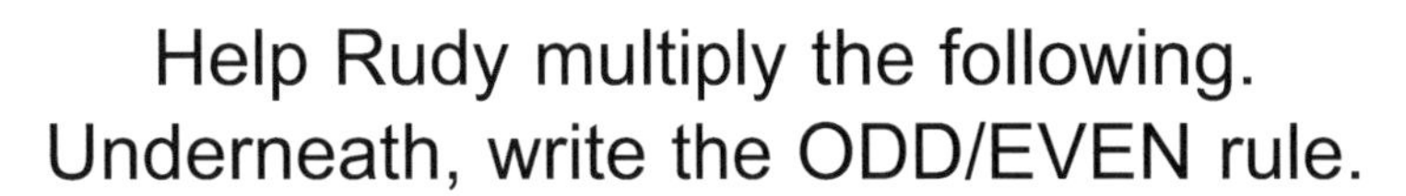

Circus Fun

Help Rudy multiply the following.
Underneath, write the ODD/EVEN rule.

6 x 4 = ____	3 x 3 = ____	8 x ___ = 48
e x e = EVEN	___ x ___ = ____	___ x ___ = ____
2 x ___ = 12	6 x 3 = ____	5 x ___ = 30
___ x ___ = ____	___ x ___ = ____	___ x ___ = ____
6 x 7 = ____	3 x 10 = ____	9 x ___ = 54
___ x ___ = ____	___ x ___ = ____	___ x ___ = ____
7 x ___ = 63	9 x ___ = 81	3 x ___ = 15
___ x ___ = ____	___ x ___ = ____	___ x ___ = ____
9 x ___ = 63	3 x ___ = 21	5 x ___ = 45
___ x ___ = ____	___ x ___ = ____	___ x ___ = ____
5 x 7 = ____	4 x ___ = 12	3 x 9 = ____
___ x ___ = ____	___ x ___ = ____	___ x ___ = ____
9 x ___ = 72	3 x ___ = 24	9 x 4 = ____
___ x ___ = ____	___ x ___ = ____	___ x ___ = ____

Shy Numbers?

Circle all the numbers that appear only ONCE on the grid.
How many are there? _____

X	1	2	3	4	5	6	7	8	9	10
1	1	2	3	4	5	6	7	8	9	10
2	2	4	6	8	10	12	14	16	18	20
3	3	6	9	12	15	18	21	24	27	30
4	4	8	12	16	20	24	28	32	36	40
5	5	10	15	20	25	30	35	40	45	50
6	6	12	18	24	30	36	42	48	54	60
7	7	14	21	28	35	42	49	56	63	70
8	8	16	24	32	40	48	56	64	72	80
9	9	18	27	36	45	54	63	72	81	90
10	10	20	30	40	50	60	70	80	90	100

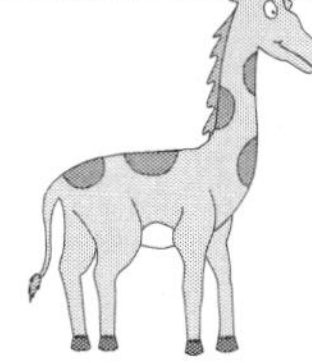

Super Easy Zero!

Help Rudy multiply the following:

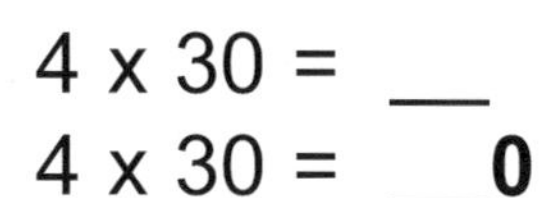

4 x 30 = __
4 x 30 = __**0**

How to solve?
Put the zero on the right.
Multiply **4 x 3.**

4 x 30 = 12**0**
4 x 30 = 120

Put **12** to the left of the zero.
Super easy, isn't it?

4 x 300 = __
4 x 300 = __**00**

How to solve?
Put the zeroes on the right.
Multiply **4 x 3**.

4 x 300 = 12**00**
4 x 300 = 1200

Put **12** to the left of the zeroes.
How easy is that?

Multiply:

3 x 80 = ____

6 x 30 = ____

9 x 70 = ____

5 x 10 = ____

90 x 2 = ____

20 x 8 = ____

50 x 7 = ____

3 x 800 = ____

6 x 300 = ____

9 x 700 = ____

5 x 100 = ____

900 x 2 = ____

200 x 8 = ____

500 x 7 = ____

Adults $5

Children $2

Joe's big brother bought 20 children's tickets. How much did he spend? $2 x 20 = $____	Mrs. Smith bought 30 adult tickets. How much did she spend? ____ x ____ = $____
Coach bought 40 children's tickets and 10 adult tickets. How much did he spend? 40 x ____ = $____ 10 x ____ = $____ Total: $____	Lisa bought 20 adult tickets and 40 children's tickets. How much did she spend? ___ x ____ = $____ ___ x ____ = $____ Total: $____
Mr. Villa spent $190 on circus tickets. He spent $40 on children's tickets. How many children's tickets did he buy? How many adult tickets did he buy? $2 x ____ = $40 Number of children: ____ $190 -$40 $150 $5 x ____ = $150 Number of adults: ____	The school spent $1100 on circus tickets. $500 was spent on adult tickets How many adults went? How many children went? $5 x____ = $500 Number of adults: ____ $1100 -$500 $600 $2 x ____ = $600 Number of children: ____

Multiplying Two Digits

First multiply the number on the right. (This is the one's place, next is the ten's.)

1			1	
28	2 x 8 = 16	→	28	2 x 2 = 4
x2	Write 6 and		x2	4 + 1 = 5
6	carry the 1.		56	is correct!

10 x7	22 x6	24 x5	18 x3
15 x5	12 x6	22 x5	16 x3
12 x9	25 x4	45 x2	61 x3
82 x4	63 x3	75 x2	42 x5

Circus Snacks

Item	Price
Hot Dogs	$3
Pretzels	$2
Cotton Candy	$1
Ice Cream	$2
Popcorn	$3
Soda	$1

Maria bought 25 pretzels at the circus for her friends. How much did she spend?

25
<u>x2</u>
$

Rico bought his baseball team 18 hot dogs. How much did he spend?

Coach bought 15 bags of popcorn and 20 sodas. How much did he spend?

Kim bought 11 ice cream bars and 10 sodas. How much did she spend?

Tony and Carla bought the class 31 hot dogs, 30 sodas and 28 ice cream bars. How much did they spend?

Julia's mom bought the soccer team 16 sodas, 15 pretzels and 18 hot dogs. How much did she spend?

Magic Circus Fun

Help Rudy solve the following:

If there are 9 giraffes, how many must Rudy put in each of the 3 rings so that each ring is the same? _____

If there are 15 monkeys, how many must Rudy put in each of the 3 rings so that each ring is the same? _____

If there are 21 lions, how many must Rudy put in each of the 3 rings so that each ring is the same? _____

Let's Do Division!

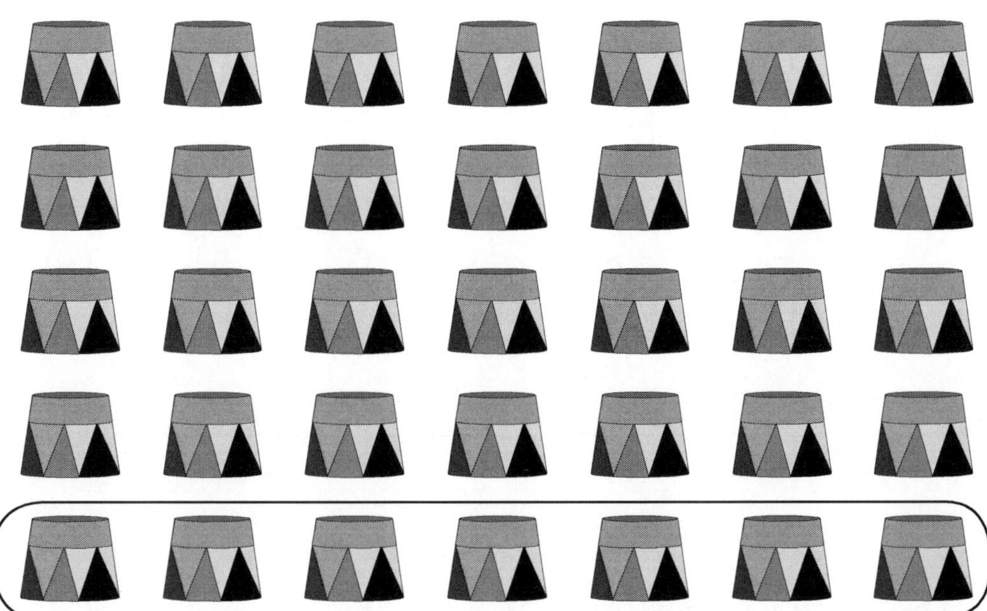

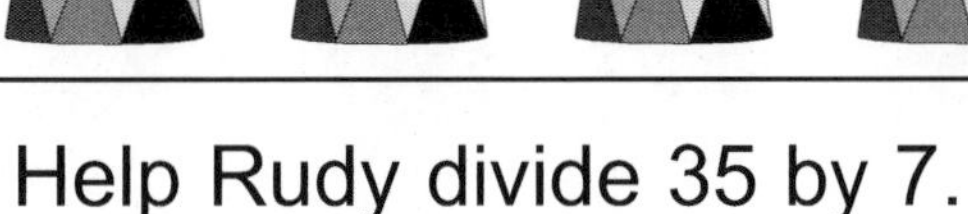

Help Rudy divide 35 by 7.

35 ÷ 7 = _____

Is the answer correct? Let's check by multiplying.

7 x _____ = 35

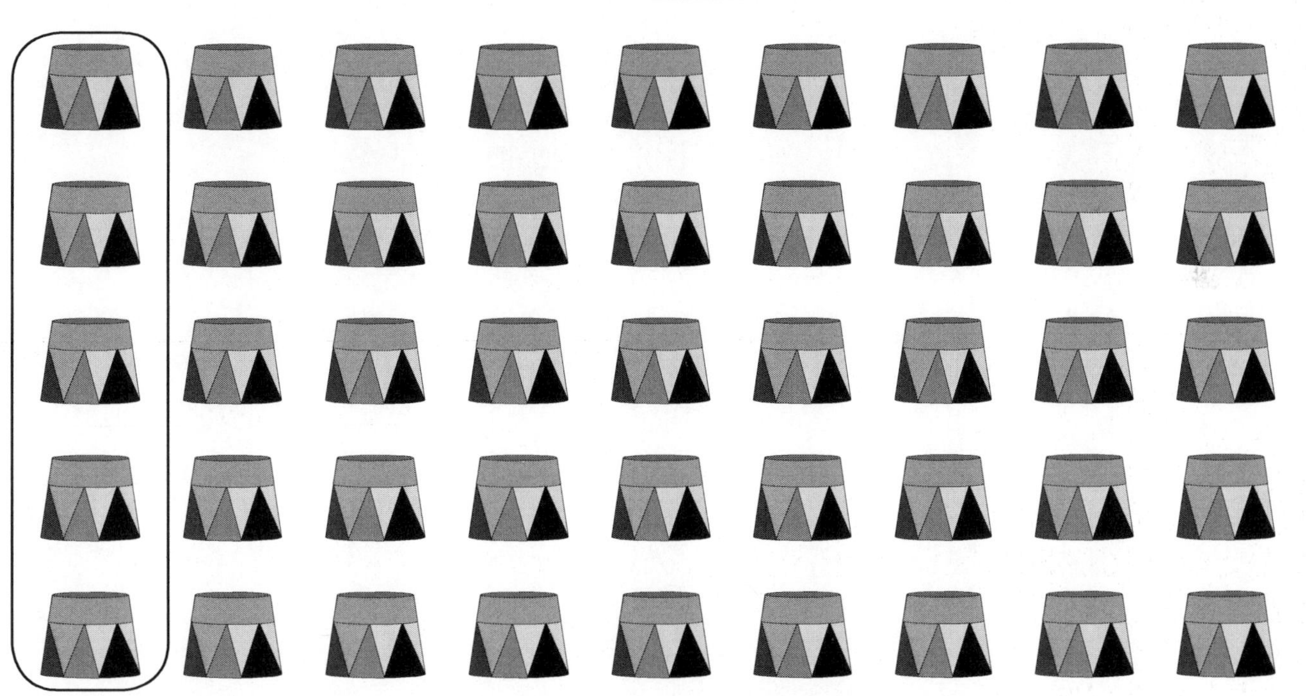

Help Rudy divide 45 by 5.

45 ÷ 5 = _____

Is the answer correct? Let's check by multiplying.

5 x _____ = 45

Patterns Are Fun!

If you divide 20 bicycles in 2 rows,
there will be _____ bicycles in each row.

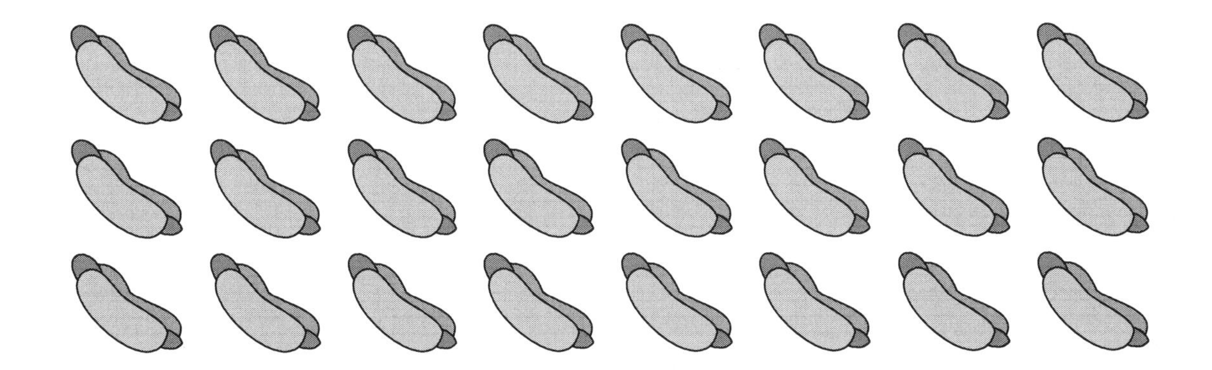

If you divide 24 hot dogs in 3 rows,
there will be _____ hot dogs in each row.

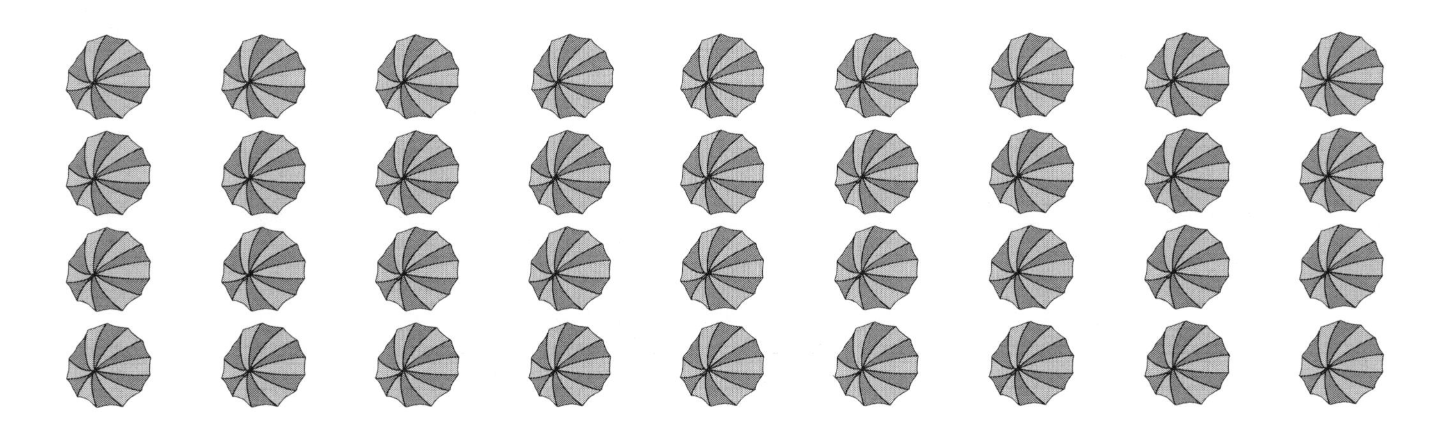

If you divide 36 umbrellas in 4 rows,
there will be _____ umbrellas in each row.

If 2 giraffes fit in a circus wagon, how many circus wagons does Rudy need for 16 giraffes? _____

If 3 monkeys fit in a circus wagon, how many circus wagons does Rudy need for 24 monkeys? _____

Help Rudy divide 15 clowns into three rings so that the same number is in each ring.

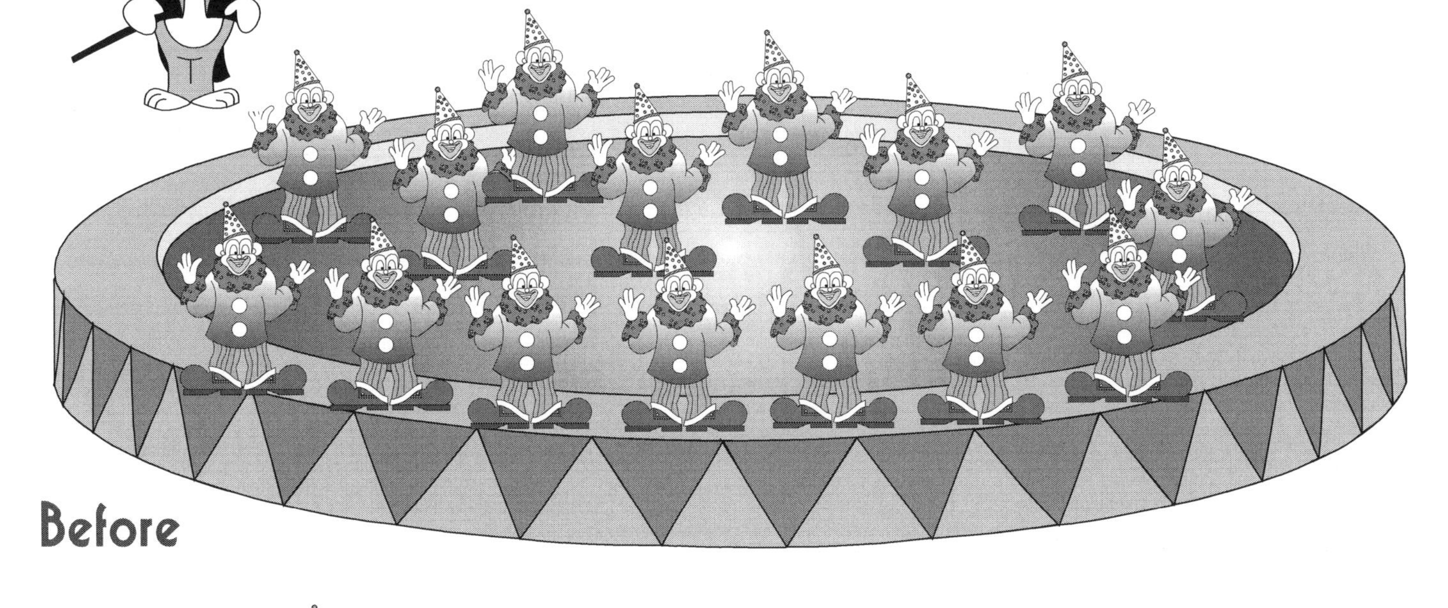

Before

After

Is this correct? If not, draw an arrow sending the extra clown to the correct ring.

$15 \div 3 =$ _____

$3 \times$ _____ $= 15$

Magic Circus Acts

Let's help Rudy with the Magic Circus.

6 clowns need to do their act. Only 2 fit in each circus wagon.
How many circus wagons does Rudy need for the clowns?

$6 \div 2 =$ _____

10 giraffes need to do their act. Only 2 fit in each circus wagon.
How many circus wagons does Rudy need for the giraffes?

$10 \div 2 =$ _____

12 monkeys need to do their act. Only 2 fit in each circus wagon.
How many circus wagons does Rudy need for the monkeys?

$12 \div 2 =$ _____

How many circus wagons will Rudy need in total?

____ + ____ + ____ = ____

Patterns Are Fun!

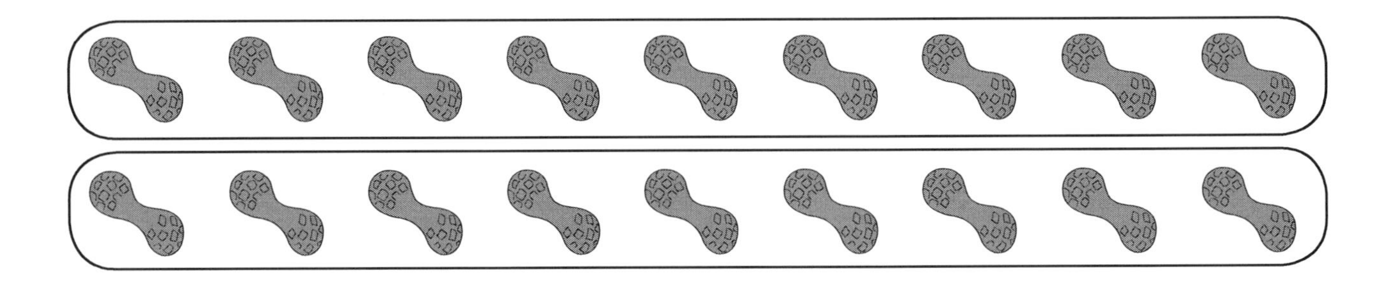

Divide 18 s into 2 groups.

18 ÷ 2 = ___ s in each group.

Divide 24 s into 8 groups.

24 ÷ 8 = ____ s in each group.

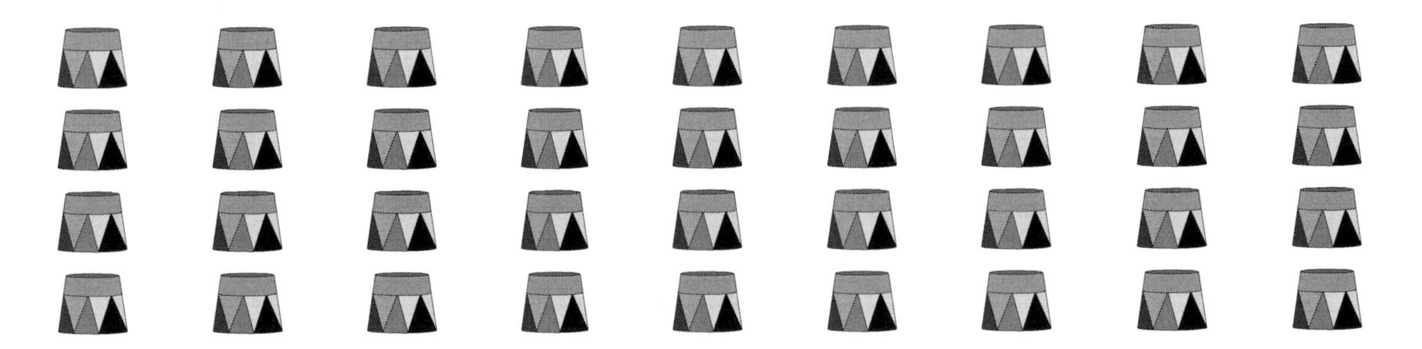

Divide 36 s into 9 groups.

36 ÷ 9 = ____ s in each group.

More Division

Division can be written: $20 \div 5 = 4$

or: $5\overline{)20}$ with quotient 4

Divide the following:

$9\overline{)18}$	$4\overline{)28}$	$7\overline{)49}$	$3\overline{)12}$
$5\overline{)45}$	$3\overline{)27}$	$6\overline{)54}$	$8\overline{)64}$
$6\overline{)30}$	$8\overline{)32}$	$5\overline{)30}$	$9\overline{)81}$
$2\overline{)16}$	$6\overline{)36}$	$8\overline{)72}$	$3\overline{)21}$
$3\overline{)24}$	$4\overline{)16}$	$5\overline{)50}$	$6\overline{)18}$
$9\overline{)36}$	$5\overline{)35}$	$7\overline{)56}$	$8\overline{)48}$

Let's Do Division!

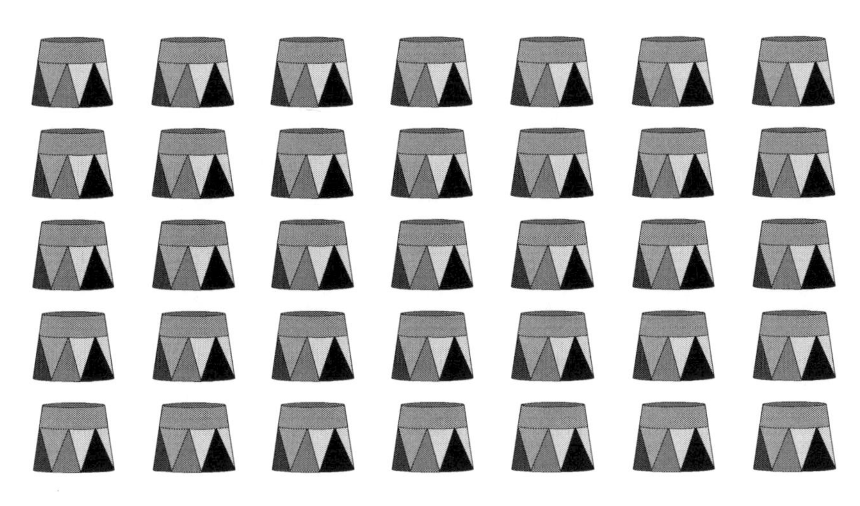

7 x 5 = 35

35 ÷ 5 = 7
or 35 ÷ 7 = 5

Turn these multiplication problems into division:

7 x 6 = 42	8 x 9 = 72	4 x 3 = 12
42 ÷ 6 = 7	___ ÷ ___ = _____	___ ÷ ___ = _____
42 ÷ 7 = 6	___ ÷ ___ = _____	___ ÷ ___ = _____
6 x 9 = 54	7 x 2 = 14	9 x 5 = 45
___ ÷ ___ = _____	___ ÷ ___ = _____	___ ÷ ___ = _____
___ ÷ ___ = _____	___ ÷ ___ = _____	___ ÷ ___ = _____
4 x 8 = 32	3 x 9 = 27	7 x 7 = 49
___ ÷ ___ = _____	___ ÷ ___ = _____	___ ÷ ___ = _____
___ ÷ ___ = _____	___ ÷ ___ = _____	___ ÷ ___ = _____

What Are the Odds?

Help Rudy solve the following.
Write two division problems for each.

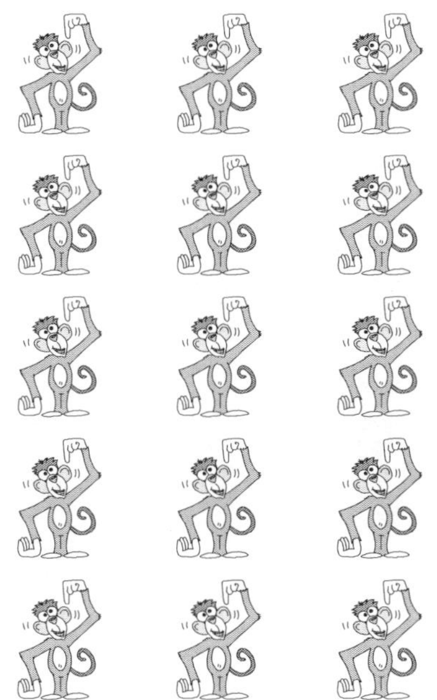

___ ÷ ___ = ___
___ ÷ ___ = ___

___ ÷ ___ = ___
___ ÷ ___ = ___

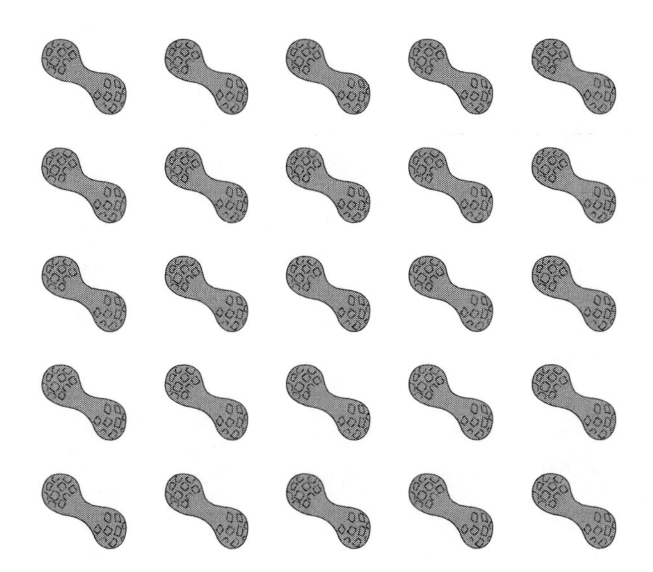

___ ÷ ___ = ___
___ ÷ ___ = ___

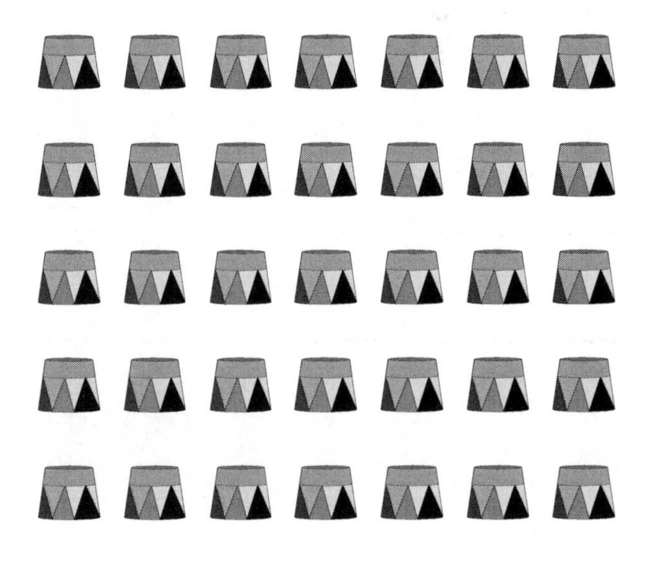

___ ÷ ___ = ___
___ ÷ ___ = ___

Fill in the blanks.

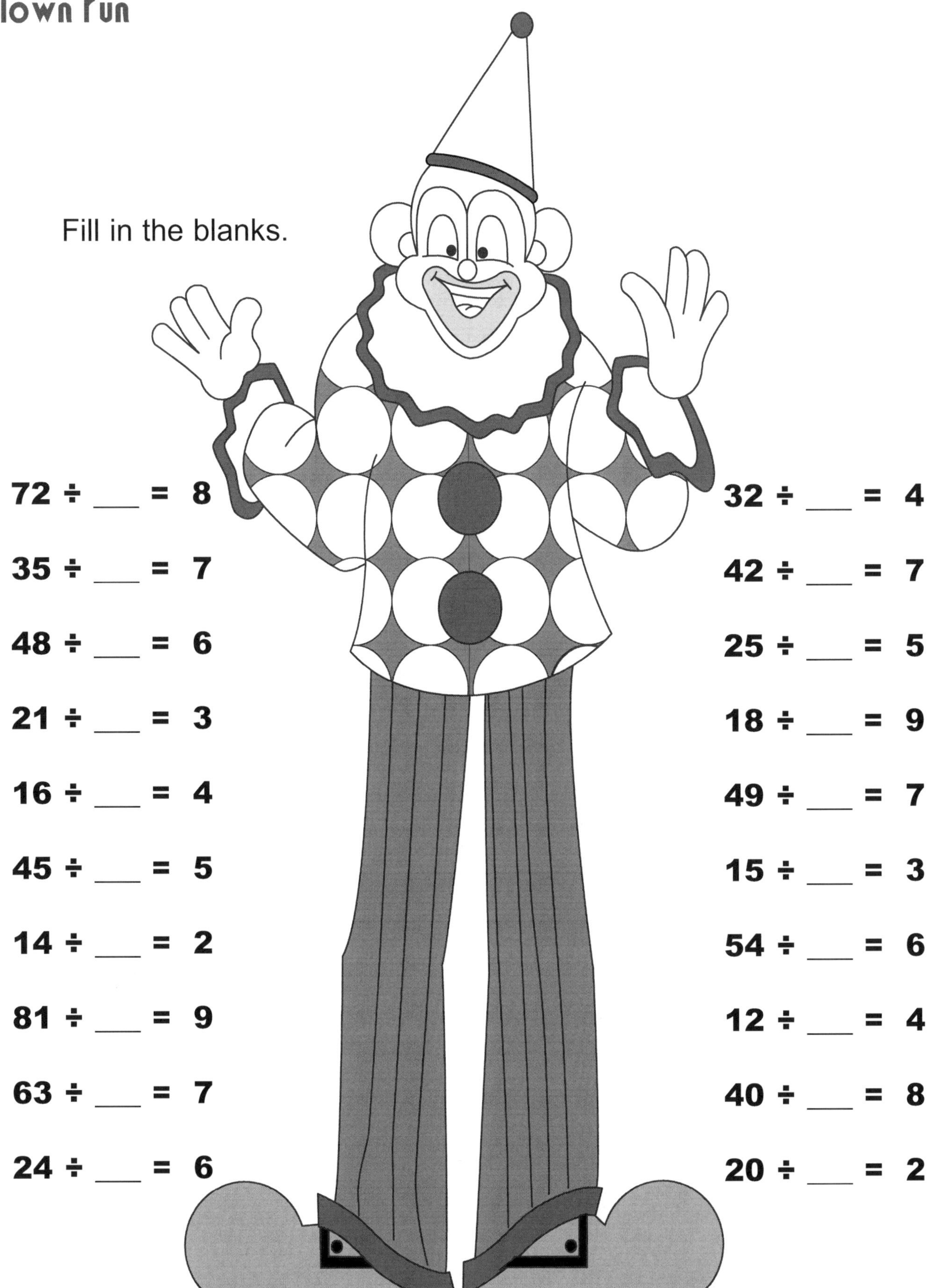

72 ÷ ___ = 8

35 ÷ ___ = 7

48 ÷ ___ = 6

21 ÷ ___ = 3

16 ÷ ___ = 4

45 ÷ ___ = 5

14 ÷ ___ = 2

81 ÷ ___ = 9

63 ÷ ___ = 7

24 ÷ ___ = 6

32 ÷ ___ = 4

42 ÷ ___ = 7

25 ÷ ___ = 5

18 ÷ ___ = 9

49 ÷ ___ = 7

15 ÷ ___ = 3

54 ÷ ___ = 6

12 ÷ ___ = 4

40 ÷ ___ = 8

20 ÷ ___ = 2

Any Remainders?

Example of division without a remainder:

```
   7
5⟌35     5 goes into 35 seven times.
 -35     7 times 5 is 35.  Subtract 35 from 35.
   0     0 left over means NO remainder.
```

Example of division with a remainder:

```
   7
5⟌38     5 goes into 38 seven times.
 -35     7 times 5 is 35.  Subtract 35 from 38.
   3     3 is left over.
```

```
   7R3
5⟌38     The answer is written 7R3.
 -35     R means remainder.
   3     Remainder means left over.
         3 was left over.
```

Circus Snacks

Hot Dogs $3
Pizza Slice $2

Ice Cream $2
Soda $1

Coach bought a large pizza with 24 slices for the team. If each of the 9 players took 2 slices, how many were left over?

$9 \times 2 =$ 18 slices $\quad$ $24 - 18 =$ 6 slices left over

Coach bought a large pizza with 24 slices for the team. If each of the 9 players took the same number of slices, how many slices were left over?

Let's solve by using division:

$$\begin{array}{r} 2 \\ 9\overline{)24} \\ \underline{-18} \\ 6 \end{array}$$

9 goes into 24 two times.

9 times 2 is 18. Subtract 18 from 24.

6 is left over.

The answer is written: 2 R6.

$$\begin{array}{r} 2\ \text{R6} \\ 9\overline{)24} \\ \underline{-18} \\ 6 \end{array}$$

R means remainder.

Remainder means left over.

6 slices were left over.

Coach bought the team 20 hot dogs. If each of the 9 players took the same number of hot dogs, how many were left over?

Check your answer. Here are 20 hot dogs.
Circle the hot dogs each of the 9 players got.
<u>Underline the hot dogs left over. These are the remainder.</u>

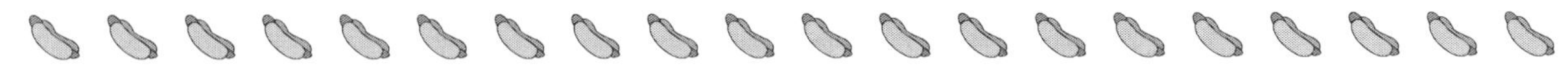

Any Remainders?

Divide the following:

$\begin{array}{r} 3 \text{ R}4 \\ 6\overline{)22} \\ -\underline{18} \\ 4 \end{array}$	$\begin{array}{r} 8 \\ 4\overline{)32} \\ -\underline{32} \\ 0 \end{array}$	$8\overline{)74}$	$3\overline{)17}$	$7\overline{)30}$
$7\overline{)49}$	$9\overline{)25}$	$4\overline{)36}$	$3\overline{)27}$	$8\overline{)39}$
$2\overline{)19}$	$9\overline{)45}$	$5\overline{)48}$	$6\overline{)38}$	$4\overline{)18}$
$7\overline{)45}$	$3\overline{)22}$	$6\overline{)36}$	$5\overline{)24}$	$9\overline{)33}$
$8\overline{)22}$	$5\overline{)47}$	$3\overline{)24}$	$6\overline{)55}$	$4\overline{)29}$

Circus Pizza

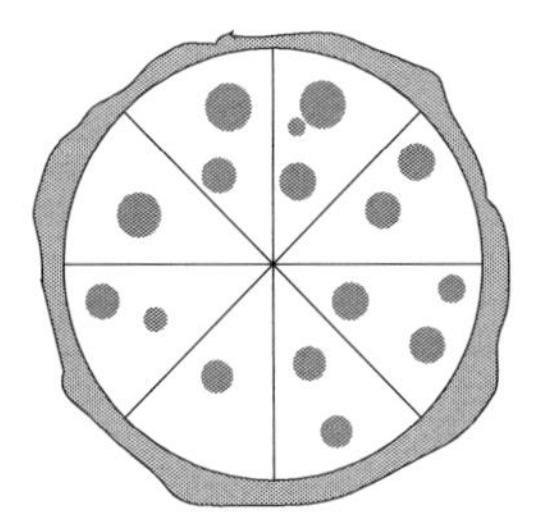 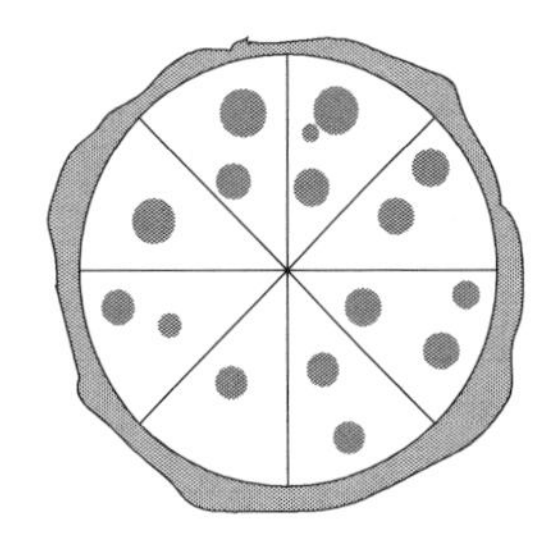 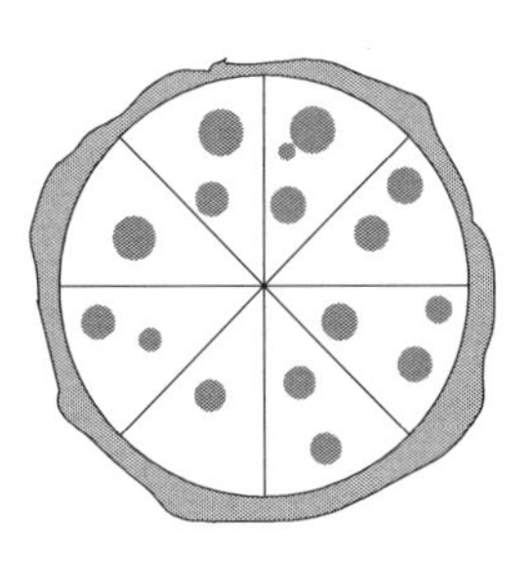

How many slices in each circus pizza? _____

How many slices altogether? ____ x ____ = _____

If coach had 3 slices, how many slices of pizza were left for the team?
(Clue: with a marker cross out 1 slice in each pizza.)

____ - ____ = ____

If the 7 team members shared the remaining slices equally, how many slices did each have?

____ ÷ ____ = ____

If the pizzas cost $12 each, how much did coach spend on 3 pizzas?

____ x _____ = $ _____

If coach bought 20 sodas at $2 each, how much did he spend on sodas?

____ x _____ = $ _____

If coach spent $19 on circus tickets, how much did he spend at the circus including pizza and sodas?

____ + ____ + ____ = $ ______

If coach handed the cashier a $100 bill, how much change did he get?

____ - ____ = $_____

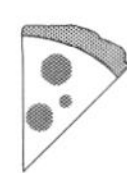  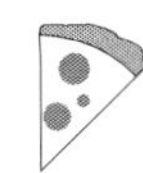  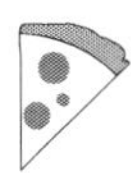 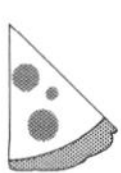 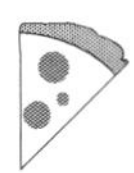 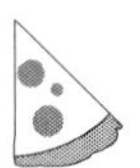

C-r-a-z-y Scramble

Fill in the grid with the correct multiple.

X	3	6	9	7	2	8	4	5	1	10
3										
6										
9										
7										
2										
8										
4										
5										
1										
10										

C-r-a-z-y Scramble
Create your own!
X

Multiple Mystery?

Not all the multiples are complete. Fill in the ______.

X	1	2	3	4	5	6	7	8	9	10
1	1	2	3	4	5	6	7	8	9	_0
2	2	4	6	8	_0	_2	_4	_6	_8	20
3	3	6	9	_2	_5	_8	21	24	27	30
4	4	8	_2	_6	20	24	28	32	36	40
5	5	_0	_5	20	25	30	35	40	45	50
6	6	_2	_8	24	30	36	42	48	54	60
7	7	_4	21	28	35	42	49	56	63	70
8	8	_6	24	32	40	48	56	64	72	80
9	9	_8	27	36	45	54	63	72	81	90
10	_0	20	30	40	50	60	70	80	90	100

Multiple Mystery?

Not all the multiples are complete.
Fill in the ______.

X	1	2	3	4	5	6	7	8	9	10
1	1	2	3	4	5	6	7	8	9	10
2	2	4	6	8	10	12	14	16	18	_0
3	3	6	9	12	15	18	_1	_4	_7	30
4	4	8	12	16	_0	_4	_8	32	36	40
5	5	10	15	_0	_5	30	35	40	45	50
6	6	12	18	_4	30	36	42	48	54	60
7	7	14	_1	_8	35	42	49	56	63	70
8	8	16	_4	32	40	48	56	64	72	80
9	9	18	_7	36	45	54	63	72	81	90
10	10	_0	30	40	50	60	70	80	90	100

Multiple Mystery?

Not all the multiples are complete.
Fill in the ______.

X	1	2	3	4	5	6	7	8	9	10
1	1	2	3	4	5	6	7	8	9	10
2	2	4	6	8	10	12	14	16	18	20
3	3	6	9	12	15	18	21	24	27	_0
4	4	8	12	16	20	24	28	_2	_6	40
5	5	10	15	20	25	_0	_5	40	45	50
6	6	12	18	24	_0	_6	42	48	54	60
7	7	14	21	28	_5	42	49	56	63	70
8	8	16	24	_2	40	48	56	64	72	80
9	9	18	27	_6	45	54	63	72	81	90
10	10	20	_0	40	50	60	70	80	90	100

Multiple Mystery?

Not all the multiples are complete.
Fill in the ______.

X	1	2	3	4	5	6	7	8	9	10
1	1	2	3	4	5	6	7	8	9	10
2	2	4	6	8	10	12	14	16	18	20
3	3	6	9	12	15	18	21	24	27	30
4	4	8	12	16	20	24	28	32	36	_0
5	5	10	15	20	25	30	35	_0	_5	50
6	6	12	18	24	30	36	_2	_8	54	60
7	7	14	21	28	35	_2	_9	56	63	70
8	8	16	24	32	_0	_8	56	64	72	80
9	9	18	27	36	_5	54	63	72	81	90
10	10	20	30	_0	50	60	70	80	90	100

Multiple Mystery?

Not all the multiples are complete.
Fill in the ______.

X	1	2	3	4	5	6	7	8	9	10
1	1	2	3	4	5	6	7	8	9	10
2	2	4	6	8	10	12	14	16	18	20
3	3	6	9	12	15	18	21	24	27	30
4	4	8	12	16	20	24	28	32	36	40
5	5	10	15	20	25	30	35	40	45	_0
6	6	12	18	24	30	36	42	48	_4	60
7	7	14	21	28	35	42	49	_6	63	70
8	8	16	24	32	40	48	_6	64	72	80
9	9	18	27	36	45	_4	63	72	81	90
10	10	20	30	40	_0	60	70	80	90	100

Multiple Mystery for 6, 7, 8 and 9!

Not all the multiples are complete.
Fill in the ______.

X	1	2	3	4	5	6	7	8	9	10
1	1	2	3	4	5	6	7	8	9	10
2	2	4	6	8	10	12	14	16	18	20
3	3	6	9	12	15	18	21	24	27	30
4	4	8	12	16	20	24	28	32	36	40
5	5	10	15	20	25	30	35	40	45	50
6	6	12	18	24	30	36	42	48	54	_0
7	7	14	21	28	35	42	49	56	_3	_0
8	8	16	24	32	40	48	56	_4	_2	_0
9	9	18	27	36	45	54	_3	_2	_1	_0
10	10	20	30	40	50	_0	_0	_0	_0	100

Multiple Mystery Review

Not all the multiples are complete.
Fill in the ______.

X	1	2	3	4	5	6	7	8	9	10
1	1	2	3	4	5	6	7	8	9	_0
2	2	4	6	8	_0	_2	_4	_6	_8	_0
3	3	6	9	_2	_5	_8	_1	_4	_7	_0
4	4	8	_2	_6	_0	_4	_8	_2	_6	_0
5	5	_0	_5	_0	_5	_0	_5	_0	_5	_0
6	6	_2	_8	_4	_0	_6	_2	_8	_4	_0
7	7	_4	_1	_8	_5	_2	_9	_6	_3	_0
8	8	_6	_4	_2	_0	_8	_6	_4	_2	_0
9	9	_8	_7	_6	_5	_4	_3	_2	_1	_0
10	_0	_0	_0	_0	_0	_0	_0	_0	_0	_0

Multiple Mystery Solved!

Not all the multiples are complete. Fill in the ______.

X	1	2	3	4	5	6	7	8	9	10
1	_	_	_	_	_	_	_	_	_	1_
2	_	_	_	_	1_	1_	1_	1_	1_	2_
3	_	_	_	1_	1_	1_	2_	2_	2_	3_
4	_	_	1_	1_	2_	2_	2_	3_	3_	4_
5	_	1_	1_	2_	2_	3_	3_	4_	4_	5_
6	_	1_	1_	2_	3_	3_	4_	4_	5_	6_
7	_	1_	2_	2_	3_	4_	4_	5_	6_	7_
8	_	1_	2_	3_	4_	4_	5_	6_	7_	8_
9	_	1_	2_	3_	4_	5_	6_	7_	8_	9_
10	1_	2_	3_	4_	5_	6_	7_	8_	9_	10_

Multiplication Bingo!

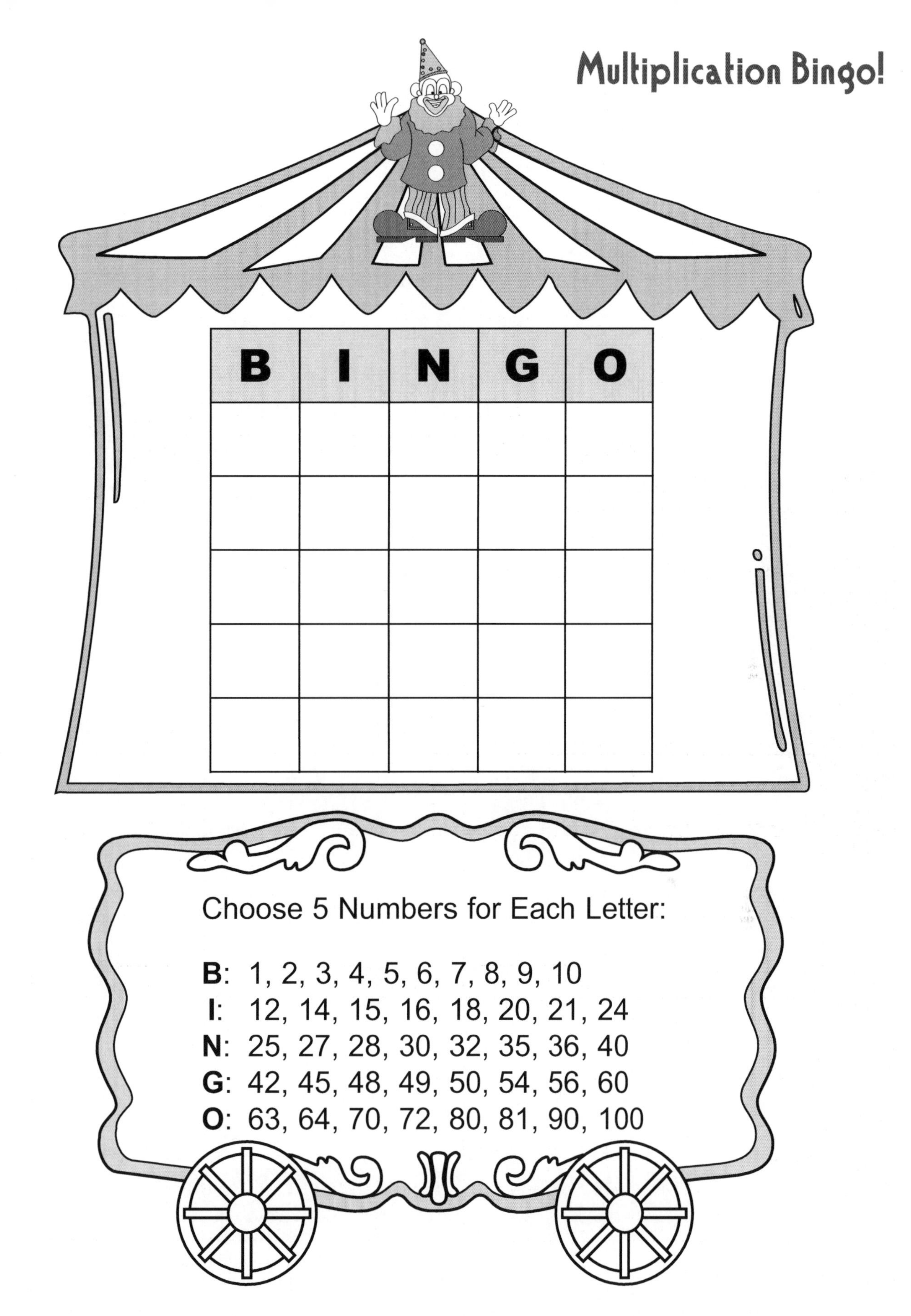

Call out multiplication problem such as: 3 x 5 = ?
Student marks the answer with an X.

ODDS and EVENS Challenge!

Fill in the grid. Circle ***all*** the ***even*** numbers. Did you notice that ***ANY*** number whether **ODD** or **EVEN** multiplied by an **EVEN** number is **EVEN**?

EVEN x ANY Number = EVEN

To reinforce the ODD/EVEN pattern, fill in ODD multiples in red and EVEN multiples in blue.

X	1	2	3	4	5	6	7	8	9	10
1										
2										
3										
4										
5										
6										
7										
8										
9										
10										

Secret Code

2 x table : 2-4-6-8 followed by **0**.
8 x table : 8-6-4-2 followed by **0**.
4 x table : 4-8-2-6 followed by **0**.
6 x table : 6-2-8-4 followed by **0**.

X	2		8
1	2		8
2	4		16
3	6		24
4	8		32
5	10		40
6	12		48
7	14		56
8	16		64
9	18		72
10	20		80

4		6
4		6
8		12
12		18
16		24
20		30
24		36
28		42
32		48
36		54
40		60

Odds and Evens Multiplication

X	1	2	3	4	5	6	7	8	9	10
1	1	2	3	4	5	6	7	8	9	10
2	2	4	6	8	10	12	14	16	18	20
3	3	6	9	12	15	18	21	24	27	30
4	4	8	12	16	20	24	28	32	36	40
5	5	10	15	20	25	30	35	40	45	50
6	6	12	18	24	30	36	42	48	54	60
7	7	14	21	28	35	42	49	56	63	70
8	8	16	24	32	40	48	56	64	72	80
9	9	18	27	36	45	54	63	72	81	90
10	10	20	30	40	50	60	70	80	90	100

X	1	2	3	4	5	6	7	8	9	10
1	1	2	3	4	5	6	7	8	9	10
2	2	4	6							
3	3	6	9							
4	4	8	12							
5	5	10	15							
6	6	12	18							
7	7	14	21							
8	8	16	24							
9	9	18	27							
10	10	20	30	40	50	60	70	80	90	100

Rudy's Magic Times Tables!

To solve a multiplication problem for numbers 1-10, find one of the numbers in the top row and the other in the column on the far left. Now run your fingers down and across from these numbers.

Your fingers will meet on the correct answer!

Try it with 6 x 4 =? Did your fingers meet at 24?
Wow, it's like magic!

X	1	2	3	4	5	6	7	8	9	10
1	1	2	3	4	5	6	7	8	9	10
2	2	4	6	8	10	12	14	16	18	20
3	3	6	9	12	15	18	21	24	27	30
4	4	8	12	16	20	24	28	32	36	40
5	5	10	15	20	25	30	35	40	45	50
6	6	12	18	24	30	36	42	48	54	60
7	7	14	21	28	35	42	49	56	63	70
8	8	16	24	32	40	48	56	64	72	80
9	9	18	27	36	45	54	63	72	81	90
10	10	20	30	40	50	60	70	80	90	100

X	1	2	3	4	5	6	7	8	9	10
1	1	2	3	4	5	6	7	8	9	10
2	2	4	6	8	10	12	14	16	18	20
3	3	6	9	12	15	18	21	24	27	30
4	4	8	12	16	20	24	28	32	36	40
5	5	10	15	20	25	30	35	40	45	50
6	6	12	18	24	30	36	42	48	54	60
7	7	14	21	28	35	42	49	56	63	70
8	8	16	24	32	40	48	56	64	72	80
9	9	18	27	36	45	54	63	72	81	90
10	10	20	30	40	50	60	70	80	90	100

Magic Circus Good-by!

Thanks to you,
everyone enjoyed the Magic Circus!
You learned the multiplication tables
and got everything in order.

Monkeys swung on the trapeze.
In the ring, lions roared for Rudy.
Giraffes pranced.
Elephants danced.
Bears tumbled.
Clowns juggled.
The circus was a great success!

Now monkeys, lions, giraffes, elephants, bears
and clowns must board the circus train.
It's time to teach another child.

All of us at the Magic Circus
hope you enjoyed learning
the multiplication tables!

Math is fun!
Pass it on.
A big Magic Circus hooray for you!

Magic Circus Diploma

____

has mastered
the MULTIPLICATION TABLES

____________________ ____________________

Date Signature

Made in the USA
San Bernardino, CA
04 October 2014